Marine Climate and Climate Change

Storms, Wind Waves and Storm Surges

Ralf Weisse and Hans von Storch

Marine Climate and Climate Change

Storms, Wind Waves and Storm Surges

Dr Ralf Weisse and Professor Dr Hans von Storch
GKSS Institute for Coastal Research
Geesthacht
Germany

SPRINGER–PRAXIS BOOKS IN ENVIRONMENTAL SCIENCES
SUBJECT *ADVISORY EDITOR*: John Mason, M.B.E., B.Sc., M.Sc., Ph.D.

ISBN 978-3-540-25316-7 Springer-Verlag Berlin Heidelberg New York

Springer is part of Springer-Science + Business Media (springer.com)

Library of Congress Control Number: 2009933600

Apart from any fair dealing for the purposes of research or private study, or criticism or review, as permitted under the Copyright, Designs and Patents Act 1988, this publication may only be reproduced, stored or transmitted, in any form or by any means, with the prior permission in writing of the publishers, or in the case of reprographic reproduction in accordance with the terms of licences issued by the Copyright Licensing Agency. Enquiries concerning reproduction outside those terms should be sent to the publishers.

© Praxis Publishing Ltd, Chichester, UK, 2010

The use of general descriptive names, registered names, trademarks, etc. in this publication does not imply, even in the absence of a specific statement, that such names are exempt from the relevant protective laws and regulations and therefore free for general use.

Cover design: Jim Wilkie
Project management: OPS Ltd, Gt Yarmouth, Norfolk, UK

Printed in Germany on acid-free paper

Contents

Preface ix

List of figures xiii

List of tables xvii

List of abbreviations and acronyms xix

1   Climate and climate variability 1
  1.1   Introduction 1
  1.2   Definition of climate 1
  1.3   The climate system 3
        1.3.1   Components of the climate system 3
        1.3.2   General circulation of the atmosphere 4
        1.3.3   General circulation of the oceans 8
  1.4   Climate variability 12
        1.4.1   Internally driven and externally forced variability 12
        1.4.2   Interplay between regional and planetary climate 16
  1.5   Summary 22
  1.6   References 23

2   Marine weather phenomena 27
  2.1   Introduction 27
  2.2   Mid-latitude storms and storm tracks 27
  2.3   Tropical cyclones 35
  2.4   Wind-generated waves 44
        2.4.1   Introduction 44
        2.4.2   Long and short-term variations of the sea state 48
        2.4.3   Freak or rogue waves 51

vi **Contents**

	2.5	Tides, storm surges, and mean sea level	57
	2.5.1	Storm surges	57
	2.5.2	Tides	66
	2.5.3	Mean sea level	69
	2.6	Summary	71
	2.7	References	73

3 Models for the marine environment 77

	3.1	Introduction	77
	3.2	Quasi-realistic modeling	79
	3.3	Climate models	82
	3.3.1	General circulation models	83
	3.3.2	Global climate models	87
	3.3.3	Regional climate models	90
	3.4	Wind wave models	96
	3.4.1	The wave spectrum	96
	3.4.2	Equation for wave energy	98
	3.4.3	Frequently used parameters to describe the sea state	102
	3.5	Tide–surge models	103
	3.5.1	Shallow-water equations	105
	3.5.2	Performance of tide–surge models	106
	3.6	Summary	107
	3.7	References	108

4 How to determine long-term changes in marine climate 113

	4.1	Introduction	113
	4.2	Problems with data quality	114
	4.2.1	Data homogeneity	115
	4.2.2	Data availability	118
	4.2.3	Operational weather analyses and other derived data sets	121
	4.3	Proxy data	124
	4.4	Global reanalyses and regional reconstructions	129
	4.4.1	Can climate trends be estimated from reanalysis data?	130
	4.4.2	The NCEP/NCAR Global Reanalysis	135
	4.4.3	Other global reanalyses	140
	4.4.4	Regional reanalyses and reconstructions	142
	4.5	Regionalization techniques	144
	4.6	Scenarios and projections	149
	4.7	Detection and attribution	151
	4.7.1	The detection problem	152
	4.7.2	The attribution problem	153
	4.7.3	Trends and detection	154
	4.8	Summary	156
	4.9	References	158

Contents vii

5 Past and future changes in wind, wave, and storm surge climates 165
  5.1   Introduction 165
  5.2   Mid-latitude cyclones and storm tracks 166
       5.2.1   Past changes and variability 166
       5.2.2   Future changes 171
  5.3   Tropical cyclones 173
       5.3.1   Past changes and variability 173
       5.3.2   Future changes 176
  5.4   Wind-generated waves 180
       5.4.1   Past changes and variability 180
       5.4.2   Future changes 184
  5.5   Tides, storm surges, and mean sea level 187
       5.5.1   Mean sea level 187
       5.5.2   Storm surges 192
       5.5.3   Tides 194
  5.6   Summary 195
  5.7   References 196

Appendices 205
  A.1   Scale analysis 205
  A.2   Geostrophic wind 207
  A.3   Geopotential height and pressure as vertical coordinates 210
  A.4   Thermal wind 211
  A.5   List of symbols 212
  A.6   References 213

Index 215

Preface

Having been involved for more than 15 years in wind, wave, and storm surge research, we have been in contact with many people having different interests in these topics. Most of them were seeking long time series of data over the often poorly sampled coastal, offshore, and ocean regions. We have collaborated with a shipyard developing RoRo and RoPax ferries operating on fixed routes. Environmental conditions such as sea states or currents to be expected on these routes during the lifetime of a vessel to them was a critical issue. Companies involved in the design and operation of offshore wind farms were interested in extreme sea states and loads to be expected at their sites, but also in the frequency and duration of fair-weather windows that can be used to mount or to maintain their equipment. Oil and gas producers were concerned about possible future changes in the wind, wave, and storm surge climate because they wish to guarantee present safety levels for their platforms and equipment in the future as well. And, of course, coastal engineers were asking for statistics of sea level extremes to be used for coastal protection planning.

Despite the different interests, long time series of wind, waves, or water levels were of primary importance to all of them. Over the oceans, such data are seldom available and often data from computer models are used instead. We have been involved in the development of such computer-generated data for many years and we know that there are numerous pitfalls in the use of these data. We feel that a thorough understanding of the basics and principles behind such data can help in successfully applying them and in improving their benefits. This book therefore provides an introduction to the climate system and to climate variability. It describes and discusses high-impact marine weather phenomena such as tropical and extra-tropical cyclones, wind waves, tides, storm surges, and mean sea level. The book further provides an overview about computer models used for simulating these phenomena, discusses their use for generating computer data that are, in many cases, used as a replacement for the often limited or missing measurements, and illustrates the numerous pitfalls one may encounter in applying such data. Eventually, we review what x **Preface**

presently (as of early 2009) is known about past and future marine climate change and variability.

In more detail, Chapter 1 provides an introduction to climate and climate variability. It explains the concepts that are fundamental for our understanding of the functioning of the system and its fluctuations. Emphasis is put on the concept of different scales and their interplay, as well as on externally and internally driven climate variability.

Chapter 2 introduces in detail the high-impact marine weather phenomena this book is about. We begin with an introduction to mid-latitude storms and storm tracks, followed by a discussion of tropical cyclones. Subsequently, we also consider the impacts caused by these phenomena; namely, wind-generated waves at the sea surface (the sea state) and storm surges. Tides and mean sea level are also included as they may contribute to extreme sea levels.

Chapter 3 provides an overview of models used for simulating the marine environment. We focus on quasi-realistic models; that is, complex computer models that describe reality as much as possible and that are used as a reality substitute. The latter means that these models are employed to derive data; namely, data that have not been measured. This is done by blending models and the few existing data. The better the existing data and the more advanced the blending techniques and the models, the better the model data. However, there is usually a number of problems involved in all models, blending techniques, and existing data.

Chapter 4 deals with problems associated with such data and present techniques to assess long-term changes in marine climate. It first describes to some extent problems related to data quality and the risk for making wrong inferences when problems are ignored. Subsequently, approaches are introduced to reduce data quality problems. These comprise the use of proxy data and reanalyses. Of course, both approaches pose new problems that are discussed in some detail. Chapter 4 is also about the techniques required to assess long-term changes in marine climate. Regionalization techniques are discussed that are used to obtain information at regional and local scales from global or large-scale data. Such techniques are needed, for instance, when local data are limited but coarse-grid or large-scale data are available. Again, different approaches exist, each with its advantages and disadvantages. Scenario and projection techniques are used when potential future changes are considered. They are used widely in marine and climate science and usually address problems of the type "What ... if ...?". For example, what will happen to storm surges in the North Sea when the statistics on North Atlantic extra-tropical cyclones change? Or, what will happen to the height of storm surges in Hamburg, Germany, if dredging in the River Elbe is continued? Chapter 4 concludes with an introduction to detection and attribution techniques. Detection refers to techniques that identify ongoing changes, while attribution is used to associate causes with the changes.

Chapter 5 provides an overview of the state of knowledge of past and future marine climate change and variability. The view presented is a snapshot as of early 2009, when this chapter was completed. Apart from specifically discussing changes in tropical and extra-tropical cyclone statistics, wind waves, and water levels, there is a

Preface xi

brief discussion on the role of consensus in science and on uncertainty assessment in climate change projections.

The book is aimed at professionals in many fields, including coastal and offshore engineers, atmospheric and oceanic researchers, and perhaps coastal and offshore planners. It will also inform and be of interest to authorities and administrations involved in marine issues. Further, the book is suitable for graduate courses on climate for atmospheric, oceanic, and environmental sciences.

ACKNOWLEDGMENTS

Starting the book was easy, arriving at the final product was not. It actually took us significantly more time than originally scheduled. We are grateful to Praxis Publishing and to Clive Horwood for their patience with us. Neil Shuttlewood applied his professional editorial skills to the project; Beate Gardeike's computer and artistic skills greatly enhanced the figures in this book. Their efforts greatly improved the quality of the final product. We are particularly grateful to Heinz Günther for his helpful comments on various drafts of the wave and wave-modeling sections. Thanks also to Nikolaus Groll for his comments on the manuscript.

Many people provided graphics and/or permission to reproduce their figures. We are particularly grateful to, in alphabetical order, Lars Bärring (Lund University), Denis Bray (GKSS Research Center), John Church (CSIRO, Australia), Ulrich Cubasch (Freie Universität Berlin), Krzysztof Fortuniak (University of Lodz), Sverre Haver (Statoil), Peter Höppe (Munich Re), Hans-Jörg Isemer (GKSS Research Center), Slava Kharin (University of Victoria), Wolfgang Koch (GKSS Research Center), Christopher Landsea (NOAA), Susanne Lehner (German Aerospace Agency), Kam-biu Liu (Louisiana State University), Uwe Mikolajewicz (Max-Planck-Institute for Meteorology), Andreas Plüß (Federal Waterways Research and Engineering Institute), Ricard Ray (NASA), Konstanze Reichert (OceanWaves GmbH), Burkhardt Rockel (GKSS Research Center), Elke Roßkamp (German Weather Service), Michaela Sickmöller, Matthias Tomczak (Flinders University), Xiaolan Wang (Environment Canada), Jörg Winterfeldt (GKSS Research Center), and Friedwart Ziemer (GKSS Research Center).

Permission to reproduce copyrighted material is greatly acknowledged and has been obtained from the American Geophysical Union, the American Meteorological Society, the European Center for Medium-range Weather Forecast, the German National Weather Service, the Intergovernmental Panel on Climate Change (IPCC)/Cambridge University Press, the National Oceanic and Atmospheric Administration (NOAA), the United Nations Environment Program (UNEP)/GRID Arendal, the World Meteorological Organization (WMO), *Nature*, Springer-Verlag, Elsevier, and Wiley–Blackwell Publishing.

Figures

1.1	Components of the climate system	color
1.2	Mean zonally averaged meridional heat transport	4
1.3	General circulation of the atmosphere emerging from an isothermal state at rest on an aqua planet	6
1.4	General circulation of the atmosphere emerging from an isothermal state at rest on a planet with realistic topography and land–sea distribution	7
1.5	Schematic diagram of the general circulation of the atmosphere	8
1.6	Schematic diagram of mean wind-driven ocean circulation at the sea surface	10
1.7	July sea surface temperature anomalies with respect to the zonal average at each latitude	10
1.8	Net freshwater flux into the Southern Ocean and mass transport through Drake Passage from an ocean model experiment driven by random freshwater fluxes	15
1.9	Response of sea surface temperatures at 51°S in a simple one-dimensional model of the Antarctic Circumpolar Current	15
1.10	Response of the simple stochastic climate model to white-noise forcing	17
1.11	The first two monthly mean air pressure anomaly distributions identified in a redundancy analysis as being most strongly linked to simultaneous variations of intra-monthly percentiles of significant wave height at Brent	20
2.1	Schematic sketch of baroclinic conditions on a constant pressure surface without and with horizontal temperature advection	29
2.2	Latitude–height cross-section of zonally averaged wind speed for Northern Hemisphere and Southern Hemisphere winter	30
2.3	Schematic diagram of mean westwind circulation at mid-latitudes	30
2.4	Circumpolar representation of 500 hPa geopotential height at 12:00 UTC, October 5, 1998	color
2.5	Example of a mid-latitude cyclone as represented in a surface weather map on February 26, 1990	32
2.6	Storm track as represented by 2-day to 6-day eddy kinetic energy at 250 hPa from ECMWF ERA-40 reanalysis	color
2.7	Track and central pressure of the Halloween Storm and Hurricane Grace	34

xiv **Figures**

2.8	Tropical cyclone frequency.	color
2.9	Schematic sketch of tropical cyclone structure	color
2.10	Sea level pressure, wind speed, and significant wave height as observed from a buoy in the Gulf of Mexico during the passage of Hurricane Kate in 1985. .	40
2.11	Wind field of Hurricane Katrina at 15:50 UTC, August 28, 2005 prior to making landfall at the U.S. coast	color
2.12	Damage caused by wind and storm surge during the Galveston Hurricane, and flooding and levee repair in New Orleans after the passage of Hurricane Katrina	43
2.13	Dependence of significant wave height and wave period on near-surface wind speed	45
2.14	Wave spectrum obtained from a wave model simulation for a location in the central North Sea	47
2.15	Meteorological conditions responsible for extreme surf heights at North Shore, Oahu (Hawaii) on December 15, 2004.	color
2.16	High surf conditions along North Shore, Oahu (Hawaii) on November 16, 2004	48
2.17	Time series of sea surface elevation measured over a 20-minute time interval and corresponding significant wave height derived from 20-minute measurements over a time period of 4 hours	49
2.18	Freak wave approaching the oil freighter *Esso Languedoc* in 1980	52
2.19	Impact of a freak wave on the Norwegian freighter *Wilstar* in 1974.	53
2.20	Time series of sea surface elevation at the Draupner platform in the North Sea	55
2.21	Freak wave detected from ERS-2 SAR in August 1996 in the Southern Ocean	56
2.22	Schematic diagram of the distribution of storm surge heights along a coastline during landfall of a tropical cyclone	59
2.23	Relative frequency of observed storm surges in Cuxhaven obtained from hourly data 1958–2002.	60
2.24	Monthly 99-percentile surge heights obtained from hourly data at Cuxhaven 1958–2002.	61
2.25	Dependence of surge heights on wind speed, fetch, and water depth	color
2.26	Global map of areas where storm surges pose a significant threat to the environment	color
2.27	Differences between observed and simulated water levels at Cuxhaven in the southern North Sea	63
2.28	Observed annual maxima high-water levels in Travemünde, Schleswig-Holstein since 1825.	65
2.29	Principal lunar tide in the world oceans.	color
2.30	Tides in the North Sea.	color
3.1	Representation of the European continent in the ECHAM global atmosphere model for different grid sizes	88
3.2	Location and intensity of the winter storm track over the North Atlantic as derived from a global climate model simulation	88
3.3	Quantitative comparison of spatial distributions of mean and 20-year return values of precipitation simulated by a series of climate models with ERA-40 reanalyses.	89
3.4	Results of a survey among climate modelers who were asked to subjectively assess the skill of atmospheric models and ocean models	90

Figures   xv

3.5	Gridded representation of Denmark and surrounding land and sea areas in a contemporary high-resolution global model, a contemporary regional model, and a regional model.	92
3.6	Example of intermittent divergence of solutions in a regional atmosphere model	94
3.7	Two different forms of wave spectra representing the same sea state	99
3.8	Principal balance of wind input, dissipation, and nonlinear wave–wave interactions and of wave energy and the overall contribution from different source terms	101
3.9	Scatter plot between observed and simulated hourly water levels, and surge levels at Borkum and Cuxhaven, two tide gauges in the German Bight	color
4.1	Number of days per decade where wind speed observations at noon exceeded 7 Beaufort or more in Hamburg, Germany	117
4.2	Frequency distribution of wind estimates reported by ships of opportunity. .	118
4.3	Surface wind field analysis of Hurricane Erin at 19:30 UTC on September 9, 2001.	119
4.4	Annual percentage of reported landfalling tropical cyclones 1900–2006 in the Atlantic U.S..	120
4.5	Number of deep cyclones detected in daily German weather maps of the North Atlantic	121
4.6	Annual number of tropical storms in east Asia (gray) and maximum decrease of central pressure within 6 hours along a tropical storm track	122
4.7	Annual mean wind speed anomalies in the North Pacific in the area of ocean weather ship P	123
4.8	Location of meteorological stations from which pressure readings have been used to estimate geostrophic wind speed in the German Bight, and time series of the 1-, 10-, and 50-percentile geostrophic wind speed derived from annual distributions of daily geostrophic wind speeds in the German Bight.	125
4.9	Percentile–percentile plot of observed daily near-surface wind speed averaged over five synoptic stations in Denmark	126
4.10	Pressure record of 1835 for Lund, Sweden, and detection performance of four different storm indices	127
4.11	Storm index for northern Europe	128
4.12	Annual mean high-water levels and linear trend at Cuxhaven, Germany and the corresponding difference between annual 99-percentile and annual mean high-water levels.	color
4.13	Annually averaged total kinetic energy from the ERA-40 reanalysis, corresponding linear trends 1958–1978 and 1979–2001, and corresponding trend 1959–2001.	131
4.14	Annually averaged anomaly correlation coefficients between 5-day forecasts and reanalysis data at the time of the forecast, for the Northern and Southern Hemispheres	132
4.15	Frequency of matched and unmatched winter storm systems in the Northern and Southern Hemispheres between the NCEP and the ERA-40 reanalyses .	133
4.16	Validation of near-surface marine wind speeds at two buoy locations representative of offshore conditions in the Atlantic and of near-shore conditions in the Mediterranean Sea	147
4.17	Skill score between collocated Quickscat wind speeds, NCEP reanalysis, and dynamically downscaled NCEP reanalysis wind speeds	color

xvi **Figures**

4.18 Emission scenarios for carbon dioxide in gigatons carbon per year and sulfur dioxide in megatons sulfur per year for different SRES scenarios. . . . . color

4.19 Estimate of scaling factors by which the response of different models to different forcing functions has to be multiplied to be in agreement with observed changes in global mean temperature . . . . color

5.1 Spatial pattern of the leading EOF mode of storm track activity derived from 300 hPa bandpass-filtered velocity variance, and corresponding principal component time series for the Northern Hemisphere, the North Atlantic, and the North Pacific 168

5.2 Changes in the number and intensity of winter (November–March) extratropical cyclones in the Northern Hemisphere in high latitudes and mid-latitudes 1959–1997 . . . . color

5.3 Storm index for northwest Europe based on geostrophic wind speed percentiles color

5.4 Northern Hemisphere and individual basin accumulated tropical cyclone energy 1981–2007 175

5.5 Change in vertical wind shear in percent per degree global warming for the Northern Hemisphere and the Southern Hemisphere tropical cyclone season color

5.6 Ensemble mean of the contribution of change in individual terms in the genesis potential index to overall GPI change in percent . . . . color

5.7 Change in genesis potential index per degree global warming for the Northern Hemisphere and the Southern Hemisphere tropical cyclone season . . . . color

5.8 Estimates of linear trends 1950–2002 in significant wave height in centimeters per decade derived from VOS data . . . . color

5.9 Ensemble mean change 2080–1990 in seasonal mean significant wave height in centimeters for the A2 emission scenario . . . . color

5.10 Same as Figure 5.9 but for the B2 scenario . . . . color

5.11 Climate change signals for long-term 99-percentile significant wave height from wave model simulations 186

5.12 Global mean sea level 1870–2006 with error estimates, linear trends over 20-year periods with error estimates and corresponding histogram . . . . color

5.13 Regional rates of sea level rise 1993–2003 in millimeters per year obtained from satellite altimetry and plotted around the global mean value for the same period color

5.14 Projections of and uncertainty in global mean sea level rise and its components between 2090–2099 and 1980–1999 for different emission scenarios . . . . color

5.15 Fingerprints of relative sea level change due to ice mass variations for Antarctica, Greenland, and mountain glaciers and ice sheets . . . . color

5.16 The distribution of tide gauge stations in the analysis of Woodworth and Blackman (2004) . . . . color

5.17 Changes in mean tidal range caused by mean sea level rise . . . . color

Tables

1.1	Characteristic anomalies of intra-monthly percentiles of significant wave height at the Brent oil field in winter as obtained in a redundancy analysis	21
2.1	Estimates of the magnitude of accelerations (forces) within a synthetic tropical cyclone at $10°$ latitude for given wind speeds and distances from the storm center	39
2.2	Examples of tropical cyclones with unusual deep core pressures compiled from various sources	41
2.3	Estimates of direct wind pressure	41
2.4	Examples of tropical cyclones with unusually high death toll or damage compiled from various sources	42
2.5	Ratio of individual wave height and significant wave height and the related exceedance probabilities determined from the Rayleigh distribution	50
2.6	Some examples of freak wave incidents	54
2.7	Estimated impact of some major storm surge events compiled from various sources	58
3.1	Comparison of zonal and meridional wind speed variance for different spatial scales	95
4.1	NCEP/NCAR reanalysis output variables and category	137
4.2	Comparison of major atmospheric reanalysis projects	141
4.3	Comparison of major ocean analysis and reanalysis projects	141
5.1	Tropical cyclone frequencies in percent of present-day values as simulated by several climate models under enhanced greenhouse gas concentrations	180
5.2	Estimates of contributions by various processes to global mean sea level budget	189
A.1	Characteristic horizontal scales, vertical scales, and time scales of some types of atmospheric motion	206
A.2	Classification of atmospheric motions according to their horizontal scale	207
A.3	Characteristic scales for synoptic-scale motions at mid-latitudes	209

Abbreviations and acronyms

3DVAR	Three-Dimensional Variational Data Assimilation
4DVAR	Four-Dimensional Variational Data Assimilation
ACC	Antarctic Circumpolar Current
ACE	Accumulated Cyclone Energy (index)
AMO	Atlantic Multidecadal Oscillation
AMS	American Meteorological Society
BALTIMOS	BALTIc MOdel System
CaRD10	California Reanalysis Downscaling at 10 km
CDAS	Climate Data Assimilation System
CERFACS	Centre de Recherche et de Formation Avancée en Calcul Scientifique
CRIEPI	Central Research Institute of Electric Power Industry
DLR	Deutsches Zentrum für Luft- und Raumfahrt (German Aerospace Center)
DNMI	Norwegian Meteorological Office
DOE	Department Of Energy
ECHAM5	Global climate model developed by the Max-Planck-Institut für Meteorologie, Hamburg, Germany
ECMWF	European Center for Medium-range Weather Forecast
EOF	Empirical Orthogonal Function
ERA	ECMWF Reanalyses
ERS	European Remote Sensing Satellite
FNOC	Fleet Numerical Operational Center
GCM	General Circulation Model
GECCO	German part of Estimating the Circulation and Climate of the Ocean effort
GFDL	Geophysical Fluid Dynamics Laboratory
GIA	Glacial Isostatic Adjustment

xx **Abbreviations and acronyms**

GKSS	GKSS Research Center
GODAS	NCEP Global Ocean Data Assimilation System
GP	Genesis Potential (tropical cyclone)
HadAM3	Hadley Centre Atmosphere Model
HAMSOM	Hamburg Shelf Ocean Model
INGV	Instituto Nazionale di Geofisica e Vulcanologia
IPCC	Intergovernmental Panel on Climate Change
ITCZ	Intertropical Convergence Zone
JCDAS	JMA Climate Data Assimilation System
JMA	Japan Meteorological Agency
LAM	Limited Area Model
MOC	Meridional Overturning Circulation
MPI	Maximum Potential Intensity
NARR	North American Regional Reanalysis
NCAR	National Center for Atmospheric Research
NCEP	National Centers for Environmental Prediction
OI	Optimal Interpolation
PDI	Power Dissipation Index
POL	Proudman Oceanographic Laboratory
RCAO	Regional Coupled Atmosphere Ocean (model)
RCM	Regional Climate Model; Regional GCM
SAR	Synthetic Aperture Radar
SLP	Sea Level Pressure
SODA	Simple Ocean Data Assimilation
SRES	Special Report on Emissions Scenarios
SSM/I	Special Sensor Microwave/Imager
SST	Sea Surface Temperature
SWAN	Simulating Waves Nearshore
TELEMAC2D	A two-dimensional hydrodynamic model
THC	Thermohaline Circulation
TRIM3D	Tidal, Residual, and Intertidal Mudflat Model
UKFOAM	U.K. Forecasting Ocean Assimilation Model
VOS	Voluntary Observing Ship
WAM	Third-generation wave model developed by a group of European wave modelers
WAMOS	Wave Monitoring System based on nautical radar
WASA	Waves and Storms in the northeast Atlantic (EU project)
WAVEWATCH	Third-generation wave model developed at NOAA/NCEP
WMO	World Meteorological Organization
XBT	eXpendable Bathy-Thermograph

1

Climate and climate variability

1.1 INTRODUCTION

In the following we describe some basic concepts that are fundamental for understanding (marine) climate and climate variability. We begin with a brief historical review on earlier and more modern concepts of climate and subsequently define *marine weather* and *marine climate* as used throughout the remainder of this book (Section 1.2). In Section 1.3 an overview of the present understanding of the global climate system is provided. In particular, the components of the climate system and the general circulations of the atmosphere and the oceans are addressed. We further show that the planetary-scale features of atmospheric circulation may be determined without any knowledge on regional details. In Section 1.4 concepts for understanding observed climate variability are discussed. Here emphasis is put on internally driven climate variability, in particular on the concept of stochastic climate models. This concept is considered to be fundamental for understanding observed climate variability ranging from several months to hundreds of years. We conclude with a discussion of the interplay between large-scale[1] climate and regional-scale climate. It is shown that there are some large-scale constraints that may be used to describe regional climate variations in terms of large-scale changes. Also, feedbacks of the regional on the large scale are discussed. It is demonstrated that in general the *details* of regional climate are unimportant, while its *statistics* indeed do matter for realistically modeling the observed climate.

1.2 DEFINITION OF CLIMATE

The word *climate* is originally deduced from the Greek *climatos* meaning *tilt* or *declination*. It refers to the fact that the average weather conditions at a specific place

---

[1] For the concept of scales see Appendix A.1.

2   **Climate and climate variability**   [Ch. 1

depend largely on the angle of incidence of incoming solar radiation. It was originally used to distinguish between the *regional* aspects of the average weather conditions; for instance, between tropical and polar climates (von Storch *et al.*, 1999). In earlier times the term *climate* was mainly associated with the near-surface atmospheric parameters that had a noticeable influence on human well-being such as near-surface temperature, humidity, or surface pressure (Humboldt, 1845). Köppen (1923) defined climate as "the average weather conditions and their evolution in time[2] at a specific place" and developed a classification scheme primarily based on latitude, surface temperature, and precipitation.

In the last few decades the concept of climate has considerably broadened. It was recognized that oceans, ice sheets, and the land surface strongly interact with the overlying atmosphere and have a profound influence on average weather conditions. Nowadays they are considered therefore as components of the *climate system* (see Section 1.3). Also the fact that the climate of a specific place is not constant but itself can vary over time became more and more acknowledged. By the year 2000, the American Meteorological Society (AMS) defined climate as "the slowly varying aspects of the atmosphere–hydrosphere–land surface system" (Glickman, 2000). The definition of the Intergovernmental Panel on Climate Change (IPCC) reads: "Climate in a narrow sense is usually defined as the 'average weather', or more rigorously, as the statistical description in terms of the mean and variability of relevant quantities over a period of time ranging from months to thousands or millions of years. The classical period is 30 years, as defined by the World Meteorological Organization (WMO). These quantities are most often surface variables such as temperature, precipitation, and wind. Climate in a wider sense is the state, including a statistical description, of the climate system" (Solomon *et al.*, 2007).

Within this book we are engaged with the *marine* aspects of weather and climate. While weather is often defined as "the state of *atmosphere*[3]" (Glickman, 2000), the state of the *oceans* associated with atmospheric conditions is usually not considered. Within this book we therefore use the term *marine weather* to refer to the state of the atmosphere and the corresponding state of the oceans. The term *marine climate* is defined as the statistical description of marine weather in terms of mean and variability, following the IPCC's concept for the definition of climate.

In its broader sense marine climate covers many aspects such as the statistics of ocean temperature and salinity, ocean circulation, or atmospheric conditions. This book is about some components of marine climate, namely the statistics of phenomena such as tropical and extra-tropical marine storms,[4] wind-generated waves at the sea surface (the sea state), or storm surges. In other words, this book is about *high-*

---

[2] Here only the annual course of long-term averages is referred to, in contrast to the variability of long-term averages themselves (climate variability).

[3] Here the term state refers to past, present, or future conditions of temperature, humidity, precipitation, cloudiness, wind, etc.

[4] That is, that part of the life cycle that is occurring over the oceans; in other words, before landfall.

Sec. 1.3] The climate system 3

impact aspects of marine climate. In the following we use the terms *marine weather* and *marine climate* to refer to these high-impact aspects only.

1.3 THE CLIMATE SYSTEM

1.3.1 Components of the climate system

While for many years climate was usually associated with the state of the atmosphere, the concept has considerably changed in the recent past. Nowadays the term climate generally refers to the state of the *climate system* consisting of the atmosphere, the hydrosphere, the cryosphere, the lithosphere, and the biosphere (Glickman, 2000). Here the hydrosphere consists of all water in the liquid phase distributed over the Earth, including the oceans, lakes, rivers, and subterranean water; the cryosphere comprises ice and snow on the Earth's surface including the major ice sheets of Greenland and Antarctica, other glaciers, permafrost, and sea ice; the lithosphere refers to the solid Earth—that is, the continents and the ocean floor; while the biosphere is made of the terrestrial and the marine flora and fauna (Figure 1.1, see color section).

While in many climate studies there is still a predominant interest in the state of the atmosphere, the climate system concept now explicitly acknowledges the fact that processes within and interactions among the different climate system components may have a profound impact on the climate and hence the state of the atmosphere. For instance, incoming solar radiation is larger in the tropics than in polar regions. This differential heating needs to be balanced by a poleward meridional heat transport in order to maintain quasi-stationary climate conditions. Figure 1.2 shows that this transport is accomplished not solely by the atmosphere, but that a substantial fraction is provided also by the oceans. While the atmosphere in principle dominates in mid- and high latitudes, most of the transport in low latitudes is accomplished by the oceans.

The state of the climate system is not constant but varies as a result of externally forced and internally driven variations. The distinction between the two sources is not always very clear. Usually the phrase *externally forced* refers to all variations in the climate system that are caused by external factors which are not affected by the climatic variables themselves. This comprises changes in Earth's orbital parameters or solar variations. Climate variations caused by changes in terrestial forcings—for instance, radiative relevant variations in atmospheric composition such as those caused by major volcanic eruptions or by human activity—are usually also considered as externally forced variations. Internally driven variability may be caused by internal instabilities, interactions, and feedbacks within the climate system. We will return to these issues in more detail in Section 1.4.

The components of the climate system are characterized by different *time scales*;[5] that is, the time different components need to adopt to a perturbation differs. As a

---

[5] See Appendix A.1 for a discussion of the concept of scales.

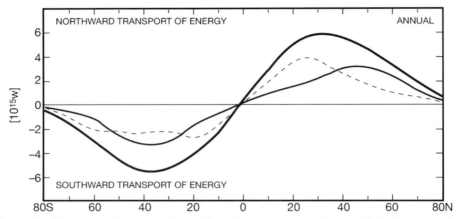

Figure 1.2. Mean zonally averaged meridional heat transport. Ocean (dashed); atmosphere (solid); ocean and atmosphere combined (heavy solid line). Redrawn after data from Peixoto and Oort (1992).

consequence the time it takes for different components to exert a noticeable influence on the climate system also varies. For instance, the lithosphere possesses the longest time scale and the slowest response within the climate system. For the phenomena and processes considered in this book such processes may therefore be treated as constant. Similarly, variations in the biosphere can be neglected. Processes in the hydrosphere, the atmosphere, and the cryosphere and their interaction are, however, essential. From these sub-systems, atmosphere and ocean are most relevant. They will be discussed in more detail next.

1.3.2 General circulation of the atmosphere

In its broadest sense the term *general circulation* refers to the *statistics* of atmospheric motions at the *planetary scale*[6] (Glickman, 2000). A simplified sketch and an intuitive approach are provided by two numerical experiments.[7] In the first experiment Fischer *et al.* (1991) used a general atmospheric circulation model[8] in which the Earth was entirely covered with water (aqua planet). Sea surface temperatures (SSTs) were prescribed such that warmer temperatures prevail near the equator and colder temperatures occurred near the poles. The initialized SSTs were zonally symmetric; in other words, there were no temperature variations along each latitude belt. At the top of the atmosphere, incoming solar radiation was prescribed. The atmosphere was

---

[6] These are large-scale motions with typical dimensions in the order of 1,000 km (see Appendix A.1).
[7] These are experiments with quasi-realistic models of the atmosphere by means of which otherwise impossible experiments can be conducted (see Chapter 3).
[8] A global atmosphere model with a typical horizontal grid spacing of about 250 km–500 km. For details see Chapter 3.

motionless (at rest) at the beginning of the experiment. When the experiment was started, the atmosphere remained mainly motionless for about the first ten days apart from some small motions near the equator (Figure 1.3). Starting at Day 10 a rapid development can be inferred during which typical features of global mean atmospheric circulation emerge from the initially motionless background: First, tropical cells and the trade wind system are visible. Another ten days later indirect Ferrel cells with westerly and northerly near-surface winds appear. Vaguely, also a polar cell can be inferred. Kinetic energy distribution (Figure 1.3c) shows that turbulent motions first show up near the equator from where they move to their observed locations at mid-latitudes within about ten days. The circulation obtained after about one or two months corresponds rather closely to the observed global mean circulation.

In an earlier experiment Washington (1968) used a similar atmosphere model but prescribed a realistic land–sea distribution and topography. Again the experiment started from an initially motionless atmosphere and was driven by incoming solar radiation at the top of the atmosphere. As in the experiment of Fischer *et al.* (1991) a realistic planetary-scale circulation emerged within a couple of days, but in addition an imprint of the land–sea distribution and the topography on the circulation can be inferred (Figure 1.4). After about 40 days or so a characteristic macro-turbulent structure in the mid-latitudes of both hemispheres has developed, while low pressure gradients prevail in the tropics.

Three lessons can be learned from these experiments:

1. *Incoming solar radiation and differential heating[9] are the main drivers of planetary-scale atmospheric circulation.* The atmosphere represents a thermodynamic system that constantly receives and re-emits energy and transforms incoming thermal energy into mechanical energy in the form of winds in the atmosphere and currents in the ocean. Eventually, the mechanical energy provided is dissipated to thermal energy by turbulent processes and re-radiated into space.

2. *Globally averaged features of atmospheric circulation defined as the arithmetic mean of many local features may be determined directly without knowledge on local details.* The basic features of planetary-scale atmospheric circulation emerge from a state at rest within a couple of days. Regional details such as topography, land–sea distribution, or land use are unimportant for these features to evolve. Also, major features of the general circulation of the atmosphere such as tropical meridional cells or mid-latitude jet streams associated with baroclinic instabilities and the formation of storms can be simulated well for a planet entirely covered with water.

3. *Land–sea distribution as well as the largest mountain ranges such as the Himalayas, the Rocky Mountains and the Andes modify the global-scale features of atmospheric circulation.* When these features are taken into account a more realistic picture of planetary-scale circulation emerges (Figure 1.4). Their impact can already be described by low-resolution general circulation models.

---

[9] Difference in incoming solar radiation between the poles and the equator.

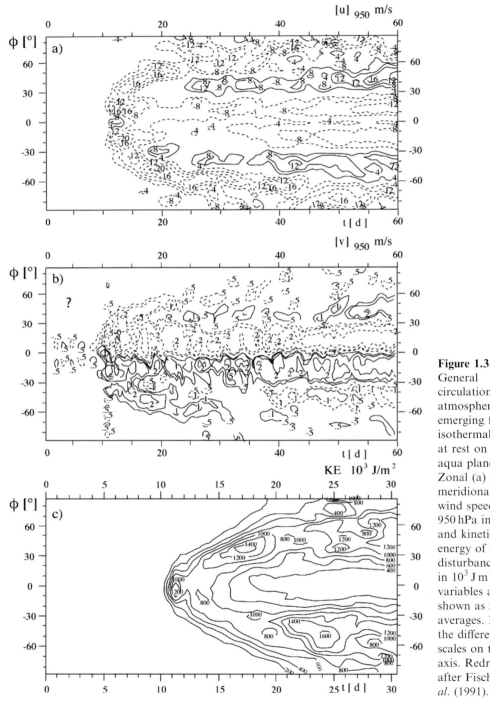

Figure 1.3. General circulation of the atmosphere emerging from an isothermal state at rest on an aqua planet. Zonal (a) and meridional (b) wind speed at 950 hPa in m s^{-1} and kinetic energy of the disturbances (c) in 10^3 J m^{-2}. All variables are shown as zonal averages. Note the different scales on the time axis. Redrawn after Fischer et al. (1991).

Figure 1.4. General circulation of the atmosphere emerging from an isothermal state at rest on a planet with realistic topography and land–sea distribution. Shown are sea surface pressure fields after 1, 5, 10, 20, and 40 days after initialization of the experiment. Redrawn after Washington (1968).

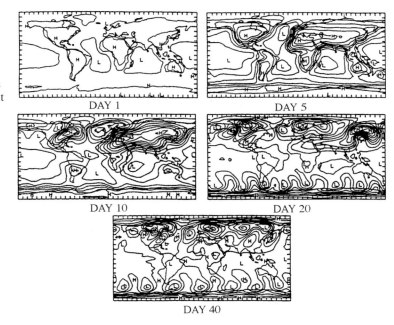

Global mean circulation as obtained from the two experiments is schematically sketched in Figure 1.5. Air warming over equatorial regions initially rises and subsequently moves poleward at higher altitudes. During its poleward propagation the air is cooling and parts of it are descending to the surface. In both hemispheres the latter occurs near 30° latitude where near-surface high-pressure systems are formed. When the descending air reaches the surface, parts of it are re-circulating equatorwards forming the so-called *trade wind* systems. The remaining part propagates poleward. Similarly, cold air is descending over the poles and propagating equatorwards. At about 60° latitude the surface branches of both air flows meet and the converging air is forced to rise.

As a result of this interplay, three meridional overturning cells can be inferred: the tropical cell is usually referred to as the *Hadley cell* while the mid-latitude cell is known as the *Ferrel cell*. The Hadley and the polar cells are both directly (thermally) driven cells, while the Ferrel cell represents a dynamical consequence of the existence of Hadley and polar cells. The surface branches of Hadley cells form the trade wind patterns. Their convergence zone is located in the tropics and is generally referred to as the *intertropical convergence zone*. Similar to the subtropics horizontal near-surface wind speeds are on average not very strong because the predominant air motion is vertical and horizontal temperature differences are small. Without horizontal temperature differences, however, there is little horizontal pressure gradient resulting in low wind conditions. The situation is different for the region where the cold surface branch of the polar cell meets the moderate temperate air provided by Ferrel cells. Here strong meridional temperature and pressure gradients occur and result in strong

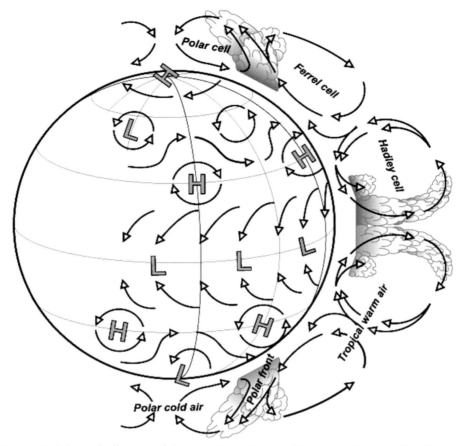

Figure 1.5. Schematic diagram of the general circulation of the atmosphere. Surface high- and low-pressure systems are indicated by the letters H and L; arrows over the globe indicate surface winds; arrows on the right-hand side indicate circulation in meridional overturning cells.

westerly winds. These regions are characterized by unsteady but statistically quasi-stationary states with short-term baroclinic disturbances; that is, synoptic weather systems such as extra-tropical low-pressure systems.

1.3.3 General circulation of the oceans

Traditionally, in oceanography circulation is divided into a wind-driven and a density-driven part. The latter was frequently referred to as thermohaline circulation (THC). Nowadays the phrase meridional overturning circulation (MOC) is used more often. The observed circulation is, however, not simply the sum of these two parts. Instead, more or less complex interactions may occur. For instance, wind-driven circulation may modify density patterns which in turn influence density-driven circulation. In general it can be stated, however, that wind-driven circulation is strongest at the surface while the density-driven part dominates at depth. In the following we briefly review mean wind and mean density-driven circulation. Also meridional heat transport provided by the ocean circulation is briefly discussed.

Wind-driven circulation. Wind-driven circulation is forced by the transfer of momentum from the atmosphere to the ocean surface. As a result, it dominates at the surface and in the upper layers of the oceans. The processes at the air–sea interface are described to some extent in Section 2.5.1. More details can be found in textbooks such as Apel (1987).

Figure 1.6 shows a schematic sketch of mean wind-driven circulation at the ocean surface. The most noticeable features are large clockwise circulation patterns in the North Atlantic and the North Pacific, as well as anticlockwise circulation patterns in the Southern Hemisphere. In all cases the circulation patterns are anticyclonic and coincide with the position and structure of atmospheric subtropical high-pressure systems, although some displacement and asymmetry can be inferred. In particular, strong poleward currents occur on the western boundaries of the gyres while a weaker and broader return flow occurs on the eastern boundaries. The gyres are generally more pronounced in the Northern than in the Southern Hemisphere. In the Southern Hemisphere a strong circumpolar current system is noticeable. It is referred to as the Antarctic circumpolar current and owes its existence to the absence of continental barriers in the extra-tropical west-wind zone of the Southern Hemisphere.

The strong poleward currents on the western boundaries are called *western boundary currents*. In the Northern Hemisphere they comprise the Gulf Stream in the North Atlantic and the Kuroshio in the North Pacific. The Brazil current and the Agulhas stream represent their less pronounced counterparts in the Southern Hemisphere. Eastern boundary currents are the California and the Peru current in the Pacific, the Canary and the Benguela current in the Atlantic, and the Leeuwin current on the western side of Australia (Figure 1.6).

Western boundary currents transport warm water from the tropics to the mid-latitudes while eastern boundary currents are associated with the equatorward transport of cold water. As a consequence the currents have an effect on the regional distribution of sea surface temperatures (SSTs). They cause a zonal asymmetry with above-average SSTs[10] at the mid- to high latitudes on the eastern side of the ocean basins which are under the influence of western boundary currents, and below average SSTs in areas where eastern boundary currents prevail (Figure 1.7).

Another, but similarly important source for observed zonal temperature anomalies is atmospheric advection. In the mid-latitudes the prevailing wind direction is from the west to the east which, in winter, brings relatively mild air masses that have been in contact with a relatively warm ocean to, for instance, Europe. On the other side of the Atlantic, the U.S. East Coast is, however, mainly under the influence of relatively cold continental air masses. This leads, on average, to colder winters when compared with average conditions at the same latitudes in Europe.

---

[10] Relative to the zonally averaged SST for each latitude.

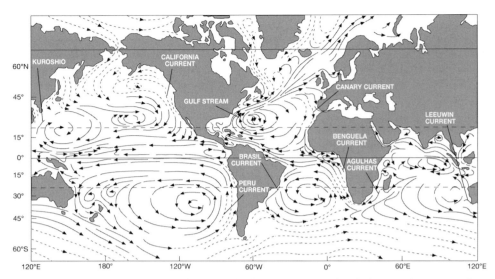

Figure 1.6. Schematic diagram of mean wind-driven ocean circulation at the sea surface. Redrawn after von Storch *et al.* (1999).

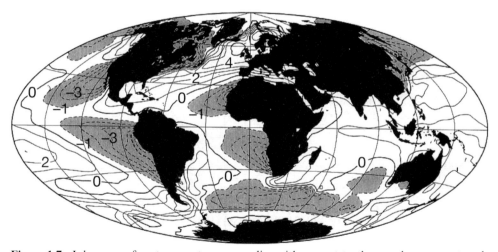

Figure 1.7. July sea surface temperature anomalies with respect to the zonal average at each latitude. The contour interval is 1°C; values less than −1°C are shaded. Redrawn after Hartmann (1994). Reproduced/modified by permission of Elsevier.

A closer inspection of Figure 1.6 reveals that, although there is some general similarity with the observed wind patterns in the atmosphere, some deviations from atmospheric patterns do occur. In particular, there is some displacement in the centers of circulation patterns and zonal asymmetry, with strong and narrow western boundary currents and broad and weaker eastern boundary currents, that is not observed in the atmosphere. An elegant explanation for this has been provided in a pioneering work by Stommel (1948). He systematically investigated the effect of a symmetrical gyral wind field on wind-driven circulation in a rectangular ocean basin. When the ocean basin was placed on a non-rotating Earth a symmetrical gyre in the ocean emerged that perfectly coincided with the atmospheric one showing no asymmetries at all. When rotation of the Earth was included in his considerations but variations of the Coriolis force with latitude were further ignored, the situation remained mostly unchanged. Although the sea surface height field appeared to be somewhat more realistic due to the onset of Ekman transport by the rotation of the Earth, the wind-driven ocean current pattern remained nearly unchanged with symmetrical stream lines following the wind field. Finally, Stommel (1948) also considered latitudinal variations of the Coriolis force. Only in this case did an ocean current field emerge with western and eastern boundary currents much as observed. This suggests that three essential factors—namely, the observed wind field, the Earth's rotation, and the latitudinal variation of the Coriolis force[11]—are responsible for the formation of the observed wind-driven circulation.

Density-driven circulation. Density differences are responsible for part of the observed circulation, in particular at depth. The density of ocean water depends on temperature and salinity. The average salinity of the oceans is about 35 psu[12] and salinity varies by less than 1 psu from ocean to ocean, top to bottom and pole to pole. Only where large rivers such as the Amazon enter the oceans and in some semi-enclosed marginal seas with freshwater excess such as the Baltic Sea, salinity may be considerably lower. Similarly, there are some restricted regions where evaporation exceeds precipitation and river runoff and where salinity may be much higher. Examples are the Red Sea and the Persian Gulf. For temperature, about 90% of the ocean volume remain perpetually between about $2°C$–$5°C$ (Hartmann, 1994). In the tropics, a much warmer surface layer (mixed layer) develops as a result of the radiation balance at the ocean surface. This mixed layer has a depth of about 300 m. At the bottom of the mixed layer a strong decrease in temperature occurs (so-called thermocline). The depth of this warm surface layer systematically decreases towards the poles. The thermocline is usually permanent in the tropics, variable in mid-latitudes where it is often deepest in summer, and shallow to nonexisting in polar regions where the water column is generally well mixed.

All oceans are characterized by permanent, well-defined density structures. These structures are a result of the unique combination of environmental conditions in each ocean basin. Water masses (layers) are formed among which there is very little mixing. For instance, Antarctic Bottom Water is formed in winter in the Weddell Sea. It represents the coldest, saltiest, and consequently the most dense water mass in the world's ocean. It fills the deepest basins in all oceans. In the Atlantic, Mediterranean water may be observed as a thin layer even south of the equator.

[11] So-called β-effect.
[12] Practical salinity units; 1 psu corresponds to 1 per mil.

12 **Climate and climate variability** [Ch. 1

Mediterranean water forms in the Eastern Mediterranean where an excess of evaporation over precipitation leads to an increase in salinity and density of the water. The water subsequently sinks to the bottom and flows toward the Western Mediterranean where it enters the Atlantic over the shallow sill of Gibraltar. In the Atlantic it still can be identified by its typical temperature and salinity.

Although generally small, density differences are responsible for a slow meridional overturning circulation in the oceans. The primary forces are pronounced cooling of the water in the northern North Atlantic and sea ice formation (which increases salinity) around Antarctica, especially in the Weddell Sea. As a result, dense water masses are formed that sink to the bottom or to intermediate depths where they subsequently produce a slow equatorward return flow. At the surface the supply of water is supported by the western boundary currents of the wind-driven circulation that transport warm water masses poleward. There are some indications that the rate of this meridional overturning may vary on time scales of decades and longer (e.g., Weisse *et al.*, 1994; Latif *et al.*, 2006).

Meridional heat transport in the oceans. The general circulation of the ocean is responsible for a meridional heat transport which is important for climate. Compared with the meridional transport provided by atmospheric circulation, oceanic transport is of the same order of magnitude, although maximum transports occur at different latitudes (Figure 1.2). Estimates for the Northern Hemisphere are about 3.2 petawatt for the ocean and 4.0 petawatt for the atmosphere (Hartmann, 1994). The numbers refer to the maxima of zonally averaged meridional heat transports and occur at about 25°N for oceanic transport and at about 45°N for atmospheric transport. An interesting question is whether wind-driven circulation, thermohaline circulation, or mid-ocean eddies generated by instabilities in zonal mean flow contribute most to oceanic meridional heat transport. A difficulty in addressing this question is that total oceanic heat transport is usually determined as a residuum between total and atmospheric transport (Hartmann, 1994). So far it seems that wind-driven and thermohaline circulation probably provide the largest contributions, at least at 20°N (Hartmann, 1994).

1.4 CLIMATE VARIABILITY

In the following we discuss concepts and relevant factors that may explain observed variability in the climate system. Subsequently the interplay between the regional-scale and planetary-scale climate is addressed; that is, the regional climate controlled by variations on the planetary scale and the feedback of regional-scale climate on large-scale climate.

1.4.1 Internally driven and externally forced variability

Climate may vary as a result of *externally forced* and *internally driven* variations (see also Section 1.3.1). The process can formally be described by a differential equation of the form

$$\frac{d}{dt} y(t) = V(y, x) + f. \tag{1.1}$$

Here $\frac{d}{dt} y(t)$ describes the behavior of a climate variable y that varies on time scales τ_y; $x(t)$ represents a climate variable that varies on a much shorter time scale τ_x; V is some nonlinear function of y and x responsible for internally driven variations; and f represents external forcing. From (1.1) it can be obtained that externally forced climate variability is induced by processes on which there is no feedback from the climate system itself. Such processes are variations in astronomical factors such as changes in the intensity of solar irradiance or changes in the Earth's orbital parameters. Terrestrial factors such as changes in atmospheric composition—for instance, caused by major volcanic eruptions or by human activity—are usually also considered as external forcing factors. While variations in astronomical factors are relatively successful in explaining variations in the Earth's climate on the time scales of ice ages, major volcanic eruptions may alter the climate on time scales of one or two years (Graf, 2002).

To account for the full spectrum of observed climate variability internally driven fluctuations have to be taken into account. There are in principle three different concepts (Hense, 2002):

1. *Non-linear interactions among the various components of the climate system.* A typical example are unstable interactions between the tropical atmosphere and the tropical ocean that are responsible for the existence of the El Niño–Southern Oscillation phenomenon. Among others, the phenomenon produces irregular fluctuations of water temperature between the date line and the South American coast.

2. *Non-linear processes.* Examples are hypotheses about the existence of different modes of meridional overturning circulation (MOC) in the oceans that are obtained from simple box models (Stommel, 1961) and have been re-examined by means of quasi-realistic ocean general circulation models (e.g., Mikolajewicz and Maier-Reimer, 1990). According to these hypotheses MOC may switch between modes of high and low meridional transport in response to surface heat and freshwater flux anomalies.

3. *Integration of short-term fluctuations by components of the climate system with a much longer response time.* The concept, known as *stochastic climate modeling*, has been introduced by Hasselmann (1976) and has been successfully applied to explain, for example, observed variations in mid-latitude sea surface temperatures (Frankignoul and Hasselmann, 1977), sea ice variability (Lemke *et al.*, 1980), or salinity and temperature fluctuations in the Antarctic circumpolar current (Weisse *et al.*, 1999).

Let us consider the last point in more detail. Prior to the introduction of this concept the general assumption was that short-term (weather) fluctuations were irrelevant for the response of the climate system that reacts slowly to such forcing. Hasselmann

14 **Climate and climate variability** [Ch. 1

(1976) noted the inconsistency of this assumption and demonstrated that low-frequency variability in the climate system may simply be the integrated response to the always present short-term weather fluctuations. To illustrate this, suppose the external forcing f in (1.1) is zero and consider the evolution of the system from an initial state y_0. For times $t > \tau_y$ the operator V may frequently be approximated by (von Storch and Zwiers, 1999)

$$V(y_t, x_t) \approx -\beta y_t + z_t. \tag{1.2}$$

where β invokes a negative feedback because the response of the system is bounded; and z_t represents a white noise[13] random process that represents short-term atmospheric fluctuations. Substitution of (1.2) into (1.1) yields

$$\frac{d}{dt} y(t) = -\beta y(t) + z(t). \tag{1.3}$$

It can be shown (e.g., Hasselmann, 1976) that the power spectrum Γ_y of this process is given by

$$\Gamma_y = \frac{\sigma_x}{\beta^2 + \omega^2} \tag{1.4}$$

where ω denotes frequency; and σ_x represents the variance of white-noise atmospheric fluctuations. The response spectrum of slowly varying components of the climate system Γ_y can deviate significantly from a white-noise process. The amplitudes of low-frequency variations are considerably larger than those at smaller frequencies and the response to short-term weather fluctuations at low frequencies can be significant. The following examples illustrate the case.

Without explicitly fitting a stochastic climate model Mikolajewicz and Maier-Reimer (1990) ran an ocean general circulation model over several thousand years. In addition to monthly mean climatologies, at the sea surface the model was forced with white-noise freshwater flux anomalies that represented short-term atmospheric fluctuations (Figure 1.8, top). The response of the ocean, in terms of mass transport through Drake passage, is dominated by variations at much lower frequencies with typical time scales of more than one hundred years (Figure 1.8, bottom). Using the same model experiment, Weisse *et al.* (1999) studied interannual fluctuations in the Antarctic circumpolar current (ACC). They found temperature and salinity anomalies that propagate along the ACC on an interannual time scale. By explicitly fitting a stochastic climate model they showed that these anomalies could be explained by the combined effects of anomaly advection with mean ocean circulation and of integration of short-term atmospheric weather fluctuations (Figure 1.9). Some similarities were found with the concept of the Antarctic Circumpolar Wave, which was proposed by White and Peterson (1996) to account for large-scale anomalies that propagate along the ACC in both the atmosphere and the ocean. Frankignoul (1995) reviewed some other applications in which dynamical systems have been modeled explicitly as stochastic climate models. These include mid-latitude sea surface tem-

---

[13] Random process with uniform spectral power density at all frequencies in the range of interest.

Sec. 1.4] **Climate variability** 15

Figure 1.8. Net freshwater flux into the Southern Ocean in $10^3 \text{ m}^3 \text{ s}^{-1}$ (top) and mass transport through Drake Passage in $10^6 \text{ m}^3 \text{ s}^{-1}$ (bottom) from an ocean model experiment driven by random (uncorrelated in space and time) freshwater fluxes. Courtesy of Uwe Mikolajewicz, Max-Planck-Institut für Meteorologie; redrawn after Mikolajewicz and Maier-Reimer (1990).

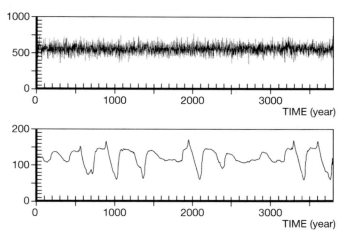

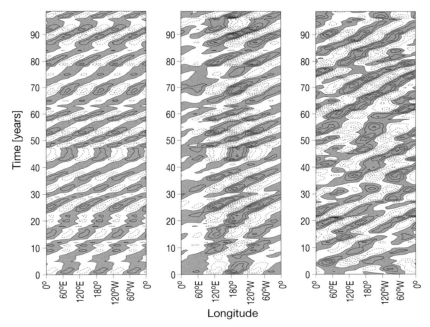

Figure 1.9. Response of sea surface temperatures at 51°S in a simple one-dimensional model of the Antarctic Circumpolar Current. Shown are the responses to a randomly forced Wavenumber 3 atmospheric pattern (left), to a superposition of randomly forced and spatially fixed Wavenumber 2 and Wavenumber 3 atmospheric patterns (middle), and to a superposition of randomly forced and spatially uncorrelated Wavenumber 2 and 3 atmospheric patterns (right). Shown are so-called Hovmoeller diagrams; that is, a cross-section plotted vs. time. The contour interval is 0.04 K; contour lines with negative values are dashed; areas with positive values are shaded. Redrawn after Weisse *et al.* (1999). Published/copyright (1999) American Geophysical Union. Reproduced/modified by permission of American Geophysical Union.

16   **Climate and climate variability**   [Ch. 1

perature (SST) anomalies and Arctic and Antarctic sea ice anomalies. For mid-latitude SSTs Frankignoul and Hasselmann (1977) have shown that the stochastic climate model successfully explains their main statistical properties, as they mainly reflect the response of the oceanic mixed layer to day-to-day changes in air–sea heat flux. For sea ice anomalies in the Arctic and Antarctic Lemke *et al.* (1980) showed that on monthly to yearly time scales these may be reasonably reflected by a stochastic climate model extended by advection. These examples efficiently demonstrate that the dynamics of the system can transform short-term weather fluctuations into low-frequency response signals that will be recognized as low-frequency climate variability.

Historically, the perception that climate may vary on time scales of decades of years and longer has repeatedly changed. Keil (1983) reports that even as recently as around 1900 there was still the notion in major meteorology textbooks that 42-year-long observational records should be sufficient to determine the *true* values of annual mean temperature and precipitation for a given location. An early pioneer who recognized that climate is not constant on longer time scales was Eduard Brückner, who proposed a 35-year quasi-periodicity (e.g., Stehr and von Storch, 2000). Nowadays, the classical period to determine a statistical description of climate in terms of *means* and *variability* is 30 years. From modern data is it obvious that even such 30-year means are not constant but may vary in time. To illustrate this let us return to the concept of internally driven climate variability as described by the stochastic climate model. The discretized version of the stochastic climate model (1.3) reads

$$y_{t+1} = (1 - \beta)y_t + z_t. \tag{1.5}$$

Assuming a damping factor (negative feedback) of $\beta = 0.2$, Figure 1.10 shows the response of the climate variable y to a white-noise forcing z with zero mean and standard deviation one. It can be inferred that (i) the response of the climate variable is dominated by variability on longer time scales and that (ii) even the 30-year (moving) averages are characterized by pronounced low-frequency variations. In the context of the ongoing discussion on anthropogenic climate change such internally induced fluctuations (natural climate variability) represent *noise* that complicates the detection and attribution of any man-made climate change (see Section 4.7 for details).

1.4.2   Interplay between regional and planetary climate

In Section 1.3.2 it was shown that the planetary-scale[14] features of atmospheric circulation can be determined without knowledge about regional details. This finding raises two questions:

1. *What determines the regional details of climate?* It is shown that regional climate variability may be understood as a combination of variations controlled and

---

[14] We will hereafter assume that the reader is familiar with the concept of scales. An introduction to the concept can be found in Appendix A.1.

Sec. 1.4] Climate variability 17

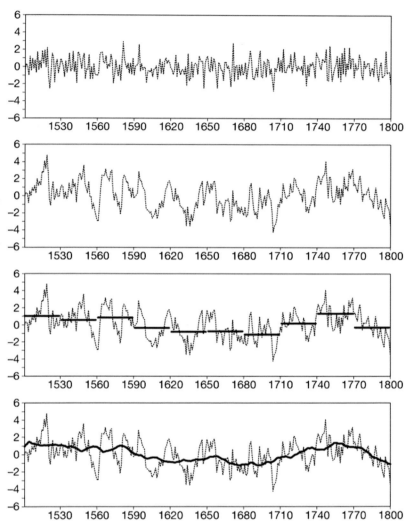

Figure 1.10. Response of the simple stochastic climate model (1.3) to white-noise (random, short-term) forcing. From top to bottom: time series of the random forcing z_t, the response of the slowly varying climate variable y_t, the response of the slowly varying climate variable y_t with 30-year averages superimposed (heavy solid line), and the response of the slowly varying climate variable y_t with 30-year moving averages superimposed. According to WMO definitions, the 30-year averages are representative of mean climate.

uncontrolled by planetary-scale climate. The contributions of these two terms vary. Often the part of variability controlled by the planetary climate is substantial.

2. *Are there feedbacks from regional climate to planetary-scale motions?* The regional aspects usually have little influence on planetary-scale climate in a sense that their *details* are unimportant. Their *statistics*, however, do matter and have a significant influence on global climate.

In the following these two questions are discussed in detail.

18   **Climate and climate variability**   [Ch. 1

Regional climate controlled by planetary scales. The relation between planetary-scale and regional climate aspects can formally be described by the following statistical approach (von Storch, 1997): Suppose that $\mathbf{G}$ represents a random variable describing the planetary or large-scale features of atmospheric circulation. Further, suppose that $\mathbf{X}$ represents a random variable that describes some regional climate aspects. Then the probability density function of the regional climate $f(\mathbf{x})$ is given by

$$f(\mathbf{x}) = \int f(\mathbf{x}, \mathbf{g}) \, d\mathbf{g} \qquad (1.6)$$

where $f(\mathbf{x}, \mathbf{g})$ represent the joint probability of $\mathbf{G}$ and $\mathbf{X}$. Note that the random variables are denoted in upper case while their realizations are denoted in lower case.

If we assume that the regional climate is conditioned upon the planetary-scale climate[15] their joint probability function can be partitioned into

$$f(\mathbf{x}, \mathbf{g}) = f(\mathbf{x}|\mathbf{g}) f(\mathbf{g}), \qquad (1.7)$$

where $f(\mathbf{x}|\mathbf{g})$ is the conditional probability function of $\mathbf{X}$ provided the random variable $\mathbf{G}$ takes the value $\mathbf{g}$. Substituting (1.7) into (1.6) yields an expression for the probability density function of regional climate

$$f(\mathbf{x}) = \int f(\mathbf{x}|\mathbf{g}) f(\mathbf{g}) \, d\mathbf{g}. \qquad (1.8)$$

It can be shown that the expectation E and the variance V of $\mathbf{X}$ can be decomposed into

$$\left. \begin{array}{l} E(\mathbf{X}) = E_{\mathbf{G}}[E_{\mathbf{X}}(\mathbf{X}|\mathbf{G})] \\[4pt] V(\mathbf{X}) = E_{\mathbf{G}}[V_{\mathbf{X}}(\mathbf{X}|\mathbf{G})] + V_{\mathbf{G}}[E_{\mathbf{X}}(\mathbf{X}|\mathbf{G})], \end{array} \right\} \qquad (1.9)$$

where the subscripts indicate the variable with respect to which the operation has been performed.

From (1.9) it can be obtained that the expectation of the regional climate is a weighted mean of conditional expectations. Further, variability of the regional climate has two contributions: The first term on the right-hand side of (1.9) represents the mean uncertainty of the conditional distribution; that is, that part of regional climate variability that is not controlled by and thus unrelated to planetary-scale climate. The second term on the right- hand side of (1.9) represents the variability of different conditional means; that is, that part of regional climate variability that is controlled by the driving planetary-scale climate.

The contribution of both terms to overall regional climate variability may vary from case to case. The smaller the first term relative to the second, the more the regional climate is controlled by planetary-scale climate. Equation (1.9) also represents the rationale for regionalization techniques usually referred to as *downscaling* (see Section 4.5). In downscaling studies local or regional observations are linked to planetary-scale climate in a statistical way. Implicitly it is assumed that the first term on the right-hand side of (1.9) is small compared with the second term in order to

---

[15] That is, regional climate is constrained by large-scale climate.

yield robust statistical relations. In other words, downscaling is applied when there are plausible arguments that a large fraction of local/regional variability can be explained by large-scale processes. The following two examples illustrate the issue.

1. *Sea breeze.* A sea breeze represents a coastal wind caused by local temperature differences. It can be observed when the sea surface is colder than the adjacent land, usually in the late afternoon on relatively calm and sunny summer days. The large-scale conditions required for this phenomenon to develop are therefore a stable surface high-pressure system and clear skies that enable substantial differential heating between the sea and the adjacent land. In terms of climate, the probability for sea breezes is higher for calm, sunny days (one realization of **G**) than for rainy and windy periods (another realization of **G**). As the probability for large-scale conditions **G** may change, so does the probability for observing sea breezes. Hence, once the planetary-scale features of atmospheric circulation are set, regional modifications appear as a result of the interplay between planetary-scale and regional aspects.

2. *Determination of local wave statistics.* The WASA-Group (1998) investigated the amount of variability in local wave (sea state) conditions that could be explained by large-scale atmospheric circulation. The analysis was performed for the Brent oil field located between Scotland and Norway. A regression model was built between the gridded monthly mean sea level pressure (SLP) field and local wave height statistics. The results are shown in Figure 1.11 and in Table 1.1. They illustrate the relation between the monthly mean SLP field and wave height intra-monthly percentiles.[16] The first large-scale monthly mean air pressure anomaly pattern shows a dipole structure with enhanced pressure gradients over the North Sea which in turn result in higher wind speeds. The pattern is associated with more extreme wave conditions at Brent (Table 1.1). Moreover, the entire probability distribution is shifted toward taller waves. The second SLP pattern describes an anomalous southerly flow across the North Sea. This pattern is associated with higher mean wave conditions (50 percentile) while extreme conditions are less severe (90 percentile, Table 1.1). Eventually, SLP fields for the period 1899–1994 and the statistical relation described above have been used to reconstruct local wave height statistics. This way the WASA-Group (1998) was able to account for 60%–70% of the observed variability in intra-monthly wave height percentiles (Table 1.1); or, in other words, 60%–70% of the observed local wave height variability could be associated with and is controlled by variations in large-scale atmospheric circulation. As with all regression models the variance in statistically derived reconstruction is smaller than the variance of the original time series. This is immediately clear as the *details* of the wave field within a month are not completely determined by the monthly mean air pressure field.

---

[16] Percentiles characterize a probability distribution. For example, the 90 percentile represents the wave height that is exceeded in only 10% of all observations while 90% remain smaller or equal that value. Intra-monthly percentiles represent the probability distribution for a month (e.g., for January).

20 Climate and climate variability [Ch. 1

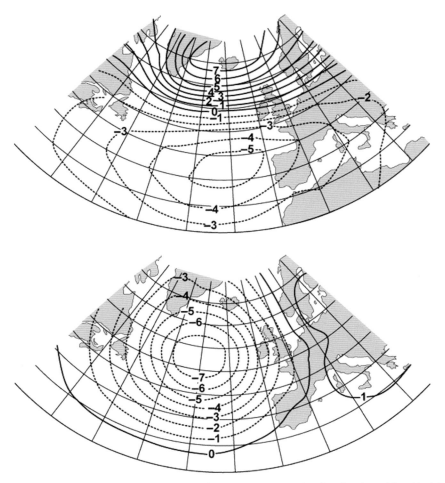

Figure 1.11. The first two monthly mean air pressure anomaly distributions identified in a redundancy analysis as being most strongly linked to simultaneous variations of intra-monthly percentiles of significant wave height at Brent (61°N, 1.5°E). The related anomalies of wave height percentiles are listed in Table 1.1. For the concept of significant wave height see Sections 2.4.1 and 3.4.3. Redrawn after WASA-Group (1998). Reproduced/modified by permission of American Meteorological Society.

The variability in local wave climate consists of an externally driven component (i.e., conditioned upon large-scale atmospheric circulation) and a component that is not or only indirectly controlled by large-scale atmospheric circulation—see Equation (1.9).

Feedback of regional climate on planetary scales. The question remains on whether or not there are feedbacks from regional climate to planetary scales; or, in other

Table 1.1. Characteristic anomalies of intra-monthly percentiles of significant wave height at the Brent oil field (61°N, 1.5°E) north of Scotland in winter (DJF) as obtained in a redundancy analysis. The k row is the kth redundancy vector. This vector represents ϵ_k of the variance of local wave height statistics within the fitting interval January 1955 until February 1995. Its coefficient shares a correlation of ρ_k with the coefficient of air pressure pattern. The last row shows the amount of local wave height variance ϵ_r that can be described by a wave height reconstruction based on the SLP for the period 1899–1994.

	Wave height (cm)				
$q =$	50%	80%	90%	ϵ_k (%)	ρ_k
$k = 1$	-86	-114	-122	94	0.84
$k = 2$	33	3	-26	5	0.08
$\epsilon_r =$	70%	67%	61%		

words, whether global climate is influenced by regional changes. To examine this, let us consider a climate variable Φ whose dynamics can in principle be described by

$$\frac{\partial \Phi}{\partial t} = \mathcal{R}(\Phi), \tag{1.10}$$

where t represents time; and $\mathcal{R}$ a suitable operator. In atmosphere general circulation models (GCMs) it is not possible to solve the full dynamics described by (1.10). Instead the equations are discretized in space and time and resolved space and time scales are *truncated* such that only large-scale processes are explicitly accounted for. If we decompose Φ into a resolved component $\overline{\Phi}$ and an unresolved part Φ'

$$\Phi = \overline{\Phi} + \Phi', \tag{1.11}$$

(1.10) is replaced by

$$\frac{\partial \overline{\Phi}}{\partial t} = \mathcal{R}(\overline{\Phi}) + \mathcal{R}'(\Phi'), \tag{1.12}$$

where the first term on the right-hand side represents the dynamics of resolved large scales while the second term summarizes the net effect of subgrid or small-scale processes on large scales. The question on whether there is a significant feedback from regional climate aspects to large-scale climate is determined by the magnitude of the second term on the right-hand side of (1.12). When the term is small, subgrid-scale processes would be irrelevant for large-scale climate. This is usually not the case. If small-scale processes are neglected in numerical models, no skillful simulation would be possible (Washington, 1999). Therefore, these processes do matter and have to be accounted for. This is usually done by conditioning them upon resolved large scales:

$$\mathcal{R}'(\Phi') \approx \mathcal{Q}(\overline{\Phi}), \tag{1.13}$$

22 Climate and climate variability [Ch. 1

where Q denotes some other suitable operator. In (1.13) the net effect of subgrid-scale processes is described as a function of resolved scales. This technique is generally known as *parameterization*. It is widely used and provides the essential backbone of all quasi-realistic models of the marine environment. Also, in the models used by Fischer *et al.* (1991) and Washington (1968) for the experiments described in Section 3.3, numerous parameterizations are included. Hence there are indeed feedbacks from small or regional-scale features to large-scale climate. However, it is the *statistics* of these processes that matters. The *details* of small-scale processes are unimportant for large-scale climate.

Some exceptions, of course, do exist. Major volcanic eruptions represent highly localized impacts. Some of them, such as Krakatoa in 1893 or Pinatubo in 1991, were that violent that they injected large amounts of aerosols into the stratosphere[17] where they did have a significant impact on planetary-scale climate (Timmreck *et al.*, 1999). Mostly, however, the influence of local events on global climate is small. For instance, Bakan *et al.* (1991) compared the impact of burning oil fields in Kuwait in 1991 in observations and model simulations. They showed that this event had only a small impact on planetary-scale circulation. Copeland *et al.* (1996) compared two atmospheric GCM experiments utilizing different land uses in the continental U.S. In the first experiment present-day farm land was assumed while in the second experiment the prairies before the conversion into farm land were prescribed. Copeland *et al.* (1996) found only small regional climate changes that were located mostly in the vicinity of the altered landscape. No planetary-scale changes were observed between the two model simulations.

Summarizing, the regional aspects of climate usually have little influence on planetary-scale climate in the sense that their *details* are unimportant. The *statistics* of regional climate of course do matter and have profound influence on global climate.

1.5 SUMMARY

While in earlier times *climate* was mainly associated with long-term averages of near-surface temperatures and precipitation, it nowadays refers to a statistical description of the state of the climate system over periods ranging from months to thousands and millions of years. In modern climatology, the climate system is considered to consist of all parts of the Earth system that may have a noticeable influence on climate; that is, the atmosphere, the hydrosphere, the cryosphere, the lithosphere, and the biosphere. Climate may vary as a result of *externally forced* and *internally driven* processes. Examples are solar variations and interactions among the different climate system components respectively. For variations ranging from several months to hundreds of years, internally driven variations are most important. An elegant concept in understanding such fluctuations has been put forward by Hasselmann (1976) with the concept of *stochastic climate models*. Here, climate

---

[17] A layer of the atmosphere at about 15 km–50 km height, just above the troposphere.

variability is explained as the integrated response of a slowly varying component of the climate system (e.g., the oceans) to the always present short-term atmospheric weather fluctuations.

The components of the climate system are characterized by highly different response times to external or internal perturbations. As a result, the time scales on which they may significantly influence the behavior of the climate system vary dramatically. Within this book we are engaged with the high-impact aspects of marine weather and climate; that is, phenomena and their statistics such as tropical and extra-tropical storms, wind waves, and storm surges. We concentrate on their recent variations over the past about one hundred years and on scenarios for the near future (50–100 years) as a result of anthropogenic climate change. On these time scales, atmosphere and ocean are the most important components of the climate system.

Both atmosphere and oceans balance the differential heating of the climate system. To do so, substantial meridional heat fluxes are required. These are provided by the atmosphere and the oceans in about equal shares. As a result, the general circulations of the atmosphere and the oceans are obtained. Using experiments with quasi-realistic climate models (see Section 1.3.2) it can be shown that the basic features of the general circulation of the atmosphere emerge from a state at rest within days and without any knowledge on regional details such as land–sea distribution or orography. When the latter are taken into account, however, a more realistic picture of general circulation evolves.

There are considerable interactions between climate at different spatial scales. Regional climate is to some extent controlled by large (planetary) scale climate. An example are local wave statistics that can to some extent be explained by anomalies in large-scale atmospheric circulation patterns (see Section 1.4.2). The latter provides the rationale for *regionalization* or *downscaling* techniques (see Section 4.5). However, not all local variance can be derived with such techniques from large-scale information. As a result, there must be parts of regional variability that cannot necessarily be attributed to variations in large-scale climate. When it comes to the question on whether or not regional climate variability matters for the large scales, it can be shown that *details* about regional variations are indeed unimportant. The *statistics* of regional climate, however, do matter (see Section 1.4.2). In climate or other models of the marine environment they are usually accounted for in the form of *parameterizations*; that is, by considering the *net effect* of subgrid or small-scale processes on the large scales. Without parameterizations, no skillful simulations would be possible.

1.6 REFERENCES

Apel, J. (1987). *Principles of Ocean Physics*. Academic Press. ISBN 0-12-058866-8.
Bakan, S.; A. Chlond; U. Cubasch; J. Feichter; H. Graf; H. Grassl; K. Hasselmann; I. Kirchner; M. Latif; E. Roeckner *et al.* (1991). Climate response to smoke from the burning oil wells in Kuwait. *Nature*, **351**, 367–371.

24 Climate and climate variability [Ch. 1]

Copeland, J.; R. Pielke; and T. Kittel (1996). Potential climatic impacts of vegetation change: A regional modeling study. *J. Geophys. Res.*, **101**, 7409–7418.

Fischer, G.; E. Kirk; and R. Podzun (1991). Physikalische Diagnose eines numerischen Experiments zur Entwicklung der großräumigen atmosphärischen Zirkulation auf einem Aquaplaneten. *Meteor. Rdsch.*, **43**, 33–42 [in German].

Frankignoul, C. (1995). Climate spectra and stochastic climate models. In: H. von Storch and A. Navarra (Eds.), *Analysis of Climate Variability: Applications of Statistical Techniques*. Springer-Verlag, pp. 29–52.

Frankignoul, C.; and K. Hasselmann (1977). Stochastic climate models. II: Application to sea-surface temperature anomalies and thermocline variability. *Tellus*, **29**, 289–305.

Glickman, T. (Ed.) (2000). *Glossary of Meteorology*. American Meteorological Society, Boston, MA, Second Edition.

Graf, H. (2002). Klimaänderungen durch Vulkane. *Promet*, **28**, 133–138 [in German]. ISSN 0340-4552.

Hartmann, D. (1994). *Global Physical Climatology*. Academic Press, London, 411 pp.

Hasselmann, K. (1976). Stochastic climate models. I: Theory. *Tellus*, **28**, 473–485.

Hense, A. (2002). Klimavariabilität durch interne Wechselwirkungen. *Promet*, **28**, 108–116 [in German]. ISSN 0340–4552.

Humboldt, A. (1845). *Kosmos: Entwurf einer physischen Weltbeschreibung*. Cotta'scher Verlag, Stuttgart, 493 pp [in German].

Keil, K. (1983). Die Entwicklung der Lehre vom Klima auf der Erde. *Die Naturwissenschaften*, **70**, 317–323 [in German].

Köppen, W. (1923). *Die Klimate der Erde*. De Gruyter, Berlin [in German].

Latif, M.; C. Bönning; J. Willebrand; A. Biastoch; J. Dengg; N. Keenlyside; and U. Schweckendiek (2006). Is the thermohaline circulation changing? *J. Climate*, **19**, 4631–4637.

Lemke, P.; W. Trinkl; and K. Hasselmann (1980). Stochastic dynamic analysis of polar sea ice variability. *J. Phys. Oceanogr.*, **10**, 2100–2120.

Mikolajewicz, U.; and E. Maier-Reimer (1990). Internal secular variability in an ocean general circulation model. *Climate Dyn.*, **4**, 145–156.

Peixoto, J.; and A. Oort (1992). *Physics of Climate*. American Institute of Physics. ISBN 0-88318-712-4.

Solomon, S.; D. Qin; M. Manning; Z. Chen; M. Marquis; K. Averyt; M. Tignor; and H. Miller (Eds.) (2007). *Climate Change 2007: The Physical Science Basis. Contribution of Working Group I to the Fourth Assessment Report of the Intergovernmental Panel on Climate Change*. Cambridge University Press, U.K. ISBN 978-0-521-88009-1, 996 pp.

Stehr, N.; and H. von Storch (Eds.) (2000). *The Sources and Consequences of Climate Change and Climate Variability in Historical Times*. Kluwer Academic. ISBN 0792361288.

Stommel, H. (1948). The westward intensification of wind-driven ocean currents. *Transactions American Geophysical Union*, **29**, 202–206.

Stommel, H. (1961). Thermohaline convection with two stable regimes. *Tellus*, **13**, 224–230.

Timmreck, C.; H.-F. Graf; and J. Feichter (1999). Simulation of Mt. Pinatubo aerosol with the Hamburg climate model ECHAM4. *Theor. and Appl. Clim.*, **62**, 85–108.

von Storch, H. (1997). *Conditional Statistical Models: A Discourse about the Local Scale in Climate Simulations*, GKSS Rep. 97/E/59. GKSS Forschungzentrum Geesthacht, Geesthacht, Germany, 58 pp.

von Storch, H.; and K. Hasselmann (1995). *Climate Variability and Change*, MPI Report 152. Max-Planck-Institut für Meteorologie, Hamburg, Germany.

Sec. 1.6] **References** 25

von Storch, H.; and F. Zwiers (1999). *Statistical Analysis in Climate Research.* Cambridge University Press, New York, 494 pp.

von Storch, H.; S. Güss; and M. Heimann (1999). *Das Klimasystem und seine Modellierung: Eine Einführung.* Springer-Verlag [in German]. ISBN 9783540658306.

WASA-Group (1998). Changing waves and storms in the Northeast Atlantic? *Bull. Am. Meteorol. Soc.,* **79**, 741–760.

Washington, W. (1968). Computer simulation of the Earth's atmosphere. *Science J.,* 37–41.

Washington, W. (1999). Three dimensional numerical simulation of climate: The fundamentals. In: H. von Storch and G. Flöser (Eds.), *Anthropogenic Climate Change.* Springer-Verlag, pp. 37–60. ISBN 3-540-65033-4.

Weisse, R.; U. Mikolajewicz; and E. Maier-Reimer (1994). Decadal variability of the North Atlantic in an ocean general circulation model. *J. Geophys. Res.,* **99**, 12411–12421.

Weisse, R.; U. Mikolajewicz; A. Sterl; and S. Drijfhout (1999). Stochastically forced variability in the Antarctic Circumpolar Current. *J. Geophys. Res.,* **104**, 11049–11064.

White, W.; and R. Peterson (1996). An Antarctic Circumpolar Wave in surface pressure, wind, temperature and sea ice extent. *Nature,* **380**, 699–702.

2

Marine weather phenomena

2.1 INTRODUCTION

In this chapter we introduce and describe some of the marine weather phenomena that may cause high impacts at sea or in coastal areas. Naturally, high wind speeds play a crucial role and they are associated with all the phenomena discussed. We start with a description of mid-latitude cyclones and storm tracks (Section 2.2). Mid-latitude cyclones form along the polar front in both hemispheres and preferably propagate eastward. The regions that, on average, experience high mid-latitude cyclone activity are referred to as storm tracks. Mid-latitude or extra-tropical cyclones are to be distinguished from tropical cyclones that preferably form over the tropical oceans within a latitude band ranging from about 5° to 20° in both hemispheres (Section 2.3). Both mid-latitude and tropical cyclones are associated with high wind speeds that are responsible for high-impact variations of sea surface height. The latter comprise wind-generated waves at the sea surface (Section 2.4) and storm surges (Section 2.5). Changes in mean sea level and tides are also addressed in Section 2.5. Although they are not related to high wind speeds, their effects may add to wind-induced variations of sea surface height and thus may significantly enhance the risk of flooding in coastal areas.

2.2 MID-LATITUDE STORMS AND STORM TRACKS

In order to understand mid-latitude storms and storm tracks we need to introduce the concepts of *barotropic* and *baroclinic* disturbances. Let us consider variation of the

28   **Marine weather phenomena**   [Ch. 2

geostrophic wind $\mathbf{v}_g$ (Appendix A.2) with height[1]

$$p\frac{\partial \mathbf{v}_g}{\partial p} = \frac{\mathbf{v}_g}{\partial(\ln p)} = -\frac{R}{f}\mathbf{k} \times \nabla_p T, \qquad (2.1)$$

where p denotes pressure; f is the Coriolis parameter; R is the gas constant; and T is temperature. For a derivation of (2.1) see Appendix A.4. $\dfrac{\partial \mathbf{v}_g}{\partial(\ln p)} = \mathbf{v}_{th}$ is also referred to as the *thermal wind*. Three principal cases can be inferred:

1. *Barotropic conditions.* These are obtained when temperatures are constant on surfaces of constant pressure;[2] that is, pressure gradients and temperature gradients are parallel such that their cross product is zero. As a result in barotropic conditions the thermal wind vector is zero and there are no variations of the geostrophic wind vector with height.
2. *Baroclinic conditions without temperature advection.* Whenever there are temperature variations on surfaces of constant pressure, baroclinic conditions are obtained. Here the thermal wind vector is different from zero; that is, the geostrophic wind vector varies with height. When the gradients of temperature and geopotential height[3] are parallel, no temperature advection is obtained. In this case there is a change in geostrophic wind speed with height, but no change in direction (Figure 2.1).
3. *Baroclinic conditions with temperature advection.* This represents the most general case. Surfaces of equal temperature and geopotential height are inclined as illustrated in the right panel of Figure 2.1. As a result both speed and direction of the geostrophic wind vector vary with height.

These three principal cases can be linked to characteristic features of the atmospheric circulation at mid-latitudes. Arranged by scale these are

1. *Mean westwind circulation at mid-latitudes.* This results from the general circulation of the atmosphere (Section 1.3.2) and occurs where the surface branches of polar and Ferrel cells converge. It can be explained under baroclinic conditions without temperature advection. Because of the existence of always present atmospheric disturbances, mean westwind circulation exists only in the long-term average.
2. *Planetary waves.* These are large wave-like disturbances of mean westwind circulation. Typically in each hemisphere there exist about two to six such dis-

---

[1] Note that in the following pressure replaces height as the vertical coordinate. The latter is possible as there exists a single-valued monotonic relation between pressure and height. In meteorology equations are often simplified when pressure is used as the vertical coordinate. For details see Appendix A.3.
[2] Strictly, there are no density variations along surfaces of constant pressure. Because of the equation of state in the atmosphere, surfaces of constant temperature are equal to surfaces of constant density. In the oceans, salinity variations have to be taken into account additionally.
[3] The height of a specified pressure surface such as the 500 hPa surface; see Appendix A.3.

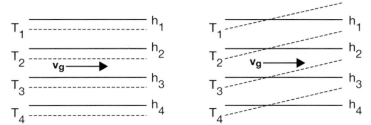

Figure 2.1. Schematic sketch of baroclinic conditions on a constant pressure surface without (left) and with (right) horizontal temperature advection. Lines of constant temperature (dashed) and geopotential height (solid).

turbances at a time, resulting in typical wavelengths of about 14,000 km to 5,000 km at 45° latitude. Planetary waves are also referred to as *barotropic Rossby waves*.

3. *Mid-latitude cyclones.* Mid-latitude or extra-tropical cyclones represent *baroclinic waves*. With typical spatial scales of about 1,000 km to 5,000 km they are considerably smaller than planetary waves. Temperature advection plays a dominant role as most of the meridional heat transport of the atmosphere at mid-latitudes is provided by mid-latitude cyclones (Hartmann, 1994).

Mean westwind circulation. Mean westwind circulation at mid-latitudes can be illustrated by the zonal mean flow. Figure 2.2 shows a latitude–height cross-section of zonally averaged wind speeds in Northern and Southern Hemisphere winter. Mean westwind circulation can be inferred from the two wind speed maxima at about 10 km height in the mid-latitudes of each hemisphere. In Southern Hemisphere winter, both maxima are shifted northwards. Further, the maximum is intensified in the Southern Hemisphere while it is reduced in the Northern Hemisphere. Figure 2.3 shows a schematic sketch for the Northern Hemisphere. Because pressure decreases more rapidly with height in cold air, pressure surfaces are inclined resulting in an increase of *geostrophic wind speed* with height. On a pressure surface, corresponding conditions are shown in the left panel of Figure 2.1.

Planetary waves. Long planetary waves are obvious features in most synoptic weather maps (Figure 2.4, see color section). Within these waves, the temperature and geopotential field are largely in phase, which means there is little temperature advection. The existence of planetary waves can be explained by *barotropic instabilities*; that is, hydrodynamic instabilities arising from a given distribution of vorticity in a two-dimensional non-divergent flow (Glickman, 2000). Consider the conservation of absolute vorticity[4] η

$$\eta = \zeta + f = \text{constant}, \tag{2.2}$$

---

[4] A vector that measures the local rotation in a fluid. Mathematically, vorticity is defined by the curl of the velocity vector.

30   Marine weather phenomena   [Ch. 2

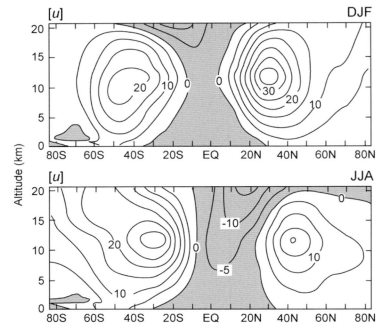

Figure 2.2. Latitude–height cross-section of zonally averaged wind speed for Northern Hemisphere (top) and Southern (bottom) Hemisphere winter. Easterly directions are shaded. Redrawn after Hartmann (1994). Copyright Elsevier (1994). Reproduced/modified by permission of Elsevier.

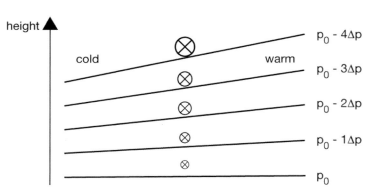

Figure 2.3. Schematic diagram of mean westwind circulation at mid-latitudes. Solid lines represent the height of the different pressure surfaces ($p_0 - n\Delta p$, $n = 0 \cdots 4$); circles with crosses illustrate the related geostrophic wind vector. Increasing diameters illustrate increasing wind speeds; the crosses denote a flow into the page.

where ζ measures relative vorticity and f denotes the Coriolis parameter that varies with latitude ϕ

$$\beta = \frac{\partial f}{\partial \phi}. \qquad (2.3)$$

Suppose, in the Northern Hemisphere, an air particle with zero absolute vorticity receives an initial disturbance such that it is deflected towards the north. Here f

increases and ζ becomes negative (anticyclonic) in order to conserve absolute vorticity. As a result, the particle is finally deflected towards the south. When it returns to its latitude of origin, the particle continues to move southward because of its inertia. As a result ζ becomes positive (cyclonic) and the particle is in turn deflected towards the north. This way planetary wave–like motions in the atmosphere are excited which are known as *barotropic Rossby waves*. The phase velocity c of these waves is given by (e.g., Pichler, 1997)

$$c = \bar{u} - \frac{\beta}{\mathbf{k}^2}.\tag{2.4}$$

The first term on the right-hand side of (2.4), $\bar{u}$, denotes the velocity of zonal mean flow. In mid-latitudes it is usually directed from west to east. The second term on the right-hand side, $-\dfrac{\beta}{\mathbf{k}^2}$, denotes a contribution that depends on β and the wavenumber $\mathbf{k}$ of the Rossby wave. It is generally directed towards the east. From (2.4) several conclusions can be inferred: First, the speed of westward propagation is limited by and generally smaller than zonal mean velocity. Second, the smaller the wavenumber (the longer the wave), the larger the contribution from the second term on the right-hand side of (2.4). Eventually, planetary waves may travel eastward. Third, there exist possible combinations of wavelength and zonal mean velocity for which planetary waves become stationary. As such, any place at mid-latitudes may remain under the influence of a certain phase of a planetary wave for several days or weeks. As it makes an essential difference whether this appears to be the warm ridge or the cold trough of the wave, the climate of mid-latitudes is to a certain extent determined by the climate of planetary waves; that is, for example, their frequency, average wavelength, or position.

Mid-latitude cyclones. Mid-latitude or extra-tropical cyclones develop at the interface between warm and cold air in mid-latitudes. They represent migratory atmospheric disturbances that are most visible in surface weather maps (Figure 2.5). Their propagation direction is generally eastward and controlled by the orientation of the flow within planetary-scale waves. Usually a cyclone with near-surface wind speeds of more than $17.2\,\mathrm{m\,s}^{-1}$ (Beaufort 8) is called a storm. Mid-latitude cyclones are characterized by *fronts*. These are relatively narrow transition zones between different air masses that are characterized by strong pressure and temperature gradients which are often associated with severe weather such as high wind speeds, heavy precipitation, or thunderstorms. While over land the potential hazards associated with mid-latitude cyclones are manifold ranging from wind damage, flooding caused by heavy precipitation, or rapid temperature decreases in combination with severe snow conditions and heavy icing (blizzards), the marine hazards associated with mid-latitude cyclones are mainly wind-related; that is, extreme wind wave conditions (see Section 2.4) or storm surges (see Section 2.5). Mid-latitude cyclones represent the dominant weather phenomenon in mid-latitudes. The process responsible for the formation of extra-tropical cyclones is referred to as *baroclinic*

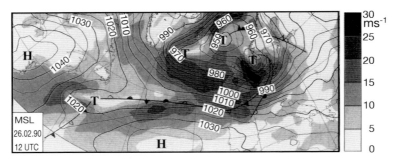

Figure 2.5. Example of a mid-latitude cyclone as represented in a surface weather map from 12:00 UTC, February 26, 1990. Contour lines represent isobars; shading illustrates near-surface wind speed. Redrawn after Kraus and Ebel (2003).

instability. Contrary to barotropic instability which is responsible for the formation of planetary-scale Rossby waves and in which only kinetic energy between the mean flow and the perturbation is transferred, in baroclinic conditions energy conversion from potential to kinetic energy plays an essential role. Baroclinic instabilities are associated with the vertical shear of the mean flow and grow by converting potential energy associated with the mean horizontal temperature gradient, which must exist to balance the thermal wind equation (2.1).

Storm tracks. A region where the synoptic[5] eddy activity is statistically and locally most prevalent and intense is generally referred to as a *storm track*. As such, the activity is mostly related to the passage of mid-latitude cyclones, and digital bandpass filters such as those proposed by Blackmon (1976) or Christoph et al. (1995) may be used to analyze the position and the intensity of storm tracks. As storm tracks roughly correspond to the mean trajectories of mid-latitude cyclones, cyclone tracking algorithms are also used to identify their positions. Figure 2.6 (see color section) provides an impression of the average location and intensity of storm tracks. In general, most of the synoptic eddy-related activity is concentrated between about 30° and 60° latitude in each hemisphere. In the Northern Hemisphere, two centers of activity can be inferred, one over the Pacific and one over the Atlantic Ocean with the latter being slightly more pronounced. They are generally referred to as the Pacific and the Atlantic storm track, respectively. In the Southern Hemisphere, the situation is zonally more symmetric, featuring only one pronounced storm track, the center of which is located over the Southern Ocean at about 50°S. In both hemispheres the storm tracks are more pronounced in winter, although the reduction in summer storm activity is larger in the Northern Hemisphere while the Southern Hemisphere storm track is present more or less throughout the year (Figure 2.6). Generally, the synoptic eddy-related variability is largest over the oceans and decreases rapidly over land. This indicates that extra-tropical cyclones play an impor-

---

[5] In meteorology *synoptic* refers to time scales of about 3–7 days.

tant role for marine weather conditions and that changes in their climatology, such as their average location or intensity, may have significant impacts on marine operations.

Regional examples of mid-latitude cyclones. Extra-tropical cyclones are phenomena affecting more or less all mid-latitude regions and mid-latitudes have adapted to their occurrence and normal intensities. From time to time, exceptional systems are observed that receive particular attention because of their unusual development or the particularly severe impacts associated with them. For the countries surrounding the North Sea, two extra-tropical wind storms affecting the region on January 31–February 1, 1953 and on February 16–17, 1962 were of particular relevance. While the 1953 storm was associated with exceptionally high northerly wind speeds over the shallow continental shelf (Wolf and Flather, 2005), the 1962 storm showed only relatively moderate wind speeds over the shelf (Koopmann, 1962). However, wind speeds were increasing over the Atlantic and an area with moderate to strong northerly winds covered a relatively large area from Northern Germany to Iceland (Müller-Navarra *et al.*, 2006). In both cases the storms caused exceptionally high storm surges and massive failure of coastal protection. For western Europe these two events represent two of the most devastating storms in recent decades with the loss of about 1,800 (Gerritsen, 2005) and 300 (Baxter, 2005) lives in The Netherlands and the U.K. during the 1953 event. The 1962 storm surge flooded large parts of the city of Hamburg in Northern Germany and caused about 300 fatalities (Sönnichsen and Moseberg, 2001). As a consequence, coastal protection strategies were adapted and massive reinforcement and remodeling of coastal protection was conducted in all three countries. For example, The Netherlands constructed barriers across several of their estuaries. In the U.K. a storm surge barrier for the River Thames was agreed upon to protect the city of London, which was eventually opened in 1984 (Wolf and Flather, 2005).

Over the Baltic Sea, Wind Storm Gudrun (January 7–9, 2005) was one of the strongest storms within the last decades. It affected most of the Nordic countries in Europe and reached hurricane strength according to the Saphir–Simpson scale. Unusual wave conditions were reported for the Baltic Proper (central Baltic Sea) with significant wave heights (see Section 2.4.2) well above 9 m, while maximum recorded wave heights before Gudrun were only 7.7 m in the northern Baltic Proper. The storm caused severe damage in northern Europe with significant cut-down of power supply in large areas of Sweden, Norway, and the Baltic states (Soomere *et al.*, 2008).

On March 12–15, 1993 a large cyclonic storm occurred over eastern North America. The massive storm complex was unusual for its size and intensity. It mainly affected the U.S. East Coast and, at its peak intensity, stretched from Canada to Central America. Near-surface wind gusts reached hurricane forces and record low barometric pressures of about 960 hPa (Cardone *et al.*, 1996) were reported. The storm was accompanied by severe whether such as tornados, thunderstorms, or snow as far south as northern Florida. Record wave heights of about 16 m were observed south of Cape Hatteras which, at this time, exceeded the existing 100-year design

34 Marine weather phenomena   [Ch. 2

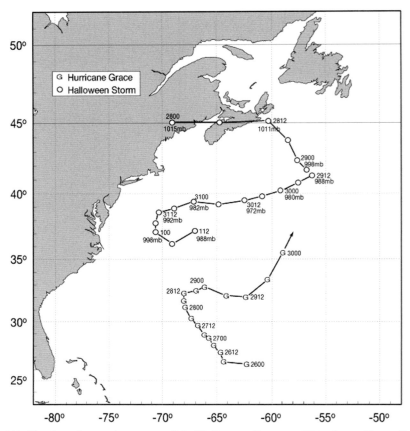

Figure 2.7. Track and central pressure of the Halloween Storm and Hurricane Grace. Redrawn after Cardone et al. (1996). Reproduced/modified by permission of American Meteorological Society.

wave height by about 50% (Cardone et al., 1996). Of the 270 casualties, 48 occurred at sea (Cardone et al., 1996).

The Perfect or Halloween Storm of October 26–November 2, 1991 was characterized by a rather unusual life cycle. A low-pressure system initially formed over the continental U.S. and moved to Atlantic Canada. On October 28 the cyclone turned abruptly southeast off Nova Scotia and rapid intensification occurred. At the same time Hurricane Grace moved slowly northwestward (Figure 2.7). Both systems began to interact and, by October 31, wind fields simplified into a pattern of extensive circulation about a single center. Henceforth the storm was referred to as the Halloween Storm (Cardone et al., 1996). Strong meridional pressure gradients of about 60 hPa over 10° latitude and maximum recorded wind speeds of about 33 m s^{-1} were observed on October 31. A Canadian deep-water buoy off Nova Scotia reported a significant wave height of 16 m (Cardone et al., 1996).

Other storms have received considerable attention because of their disastrous impacts on shipping. Examples are the extra-tropical cyclone over Bass Strait (south-east Australia) in 1998 that caused the loss of 71 sailing vessels participating in the Sydney to Hobart Yacht Race (Greenslade, 2001), or the Fastnet Storm, where a tardy meteorological warning resulted in a fleet of over 300 boats being caught in rapidly developing wind and sea conditions, only five of which eventually finished the race (van Dorn, 1994).

2.3 TROPICAL CYCLONES

Tropical cyclones are non-frontal synoptic-scale low-pressure systems over tropical or subtropical waters with organized convection (thunderstorm activity) and cyclonic surface wind circulation (Shepherd and Knutson, 2007). Originating over the tropical oceans, mature tropical cyclones typically have diameters in the order of about 100 km to 1,000 km. Pressure gradients and near-surface winds are nearly always much stronger than for extra-tropical cyclones, making the phenomenon one of the most intense storm systems of the world in terms of wind, precipitation, ocean wave heights, and storm surges. Maintained mainly by the extraction of latent heat from the oceans at high temperatures and by heat export to the upper troposphere at low temperatures, tropical cyclones represent a warm-season phenomenon that occurs only in the summer months of the respective hemisphere. Usually, activity peaks in late summer when upper-ocean temperatures are highest.

Tropical cyclones are classified according to their near-surface wind speeds and damage potential. Cyclones with wind speeds of up to $17 \, \mathrm{m \, s}^{-1}$ are referred to as tropical depressions; systems in which near-surface wind speeds reach values between $18 \, \mathrm{m \, s}^{-1}$ and $32 \, \mathrm{m \, s}^{-1}$ are called tropical storms, and cyclones with wind speeds of more than $32 \, \mathrm{m \, s}^{-1}$ are classified as severe tropical cyclones, typhoons, or hurricanes depending on region (Shepherd and Knutson, 2007). Here wind speed refers to *sustained* wind speeds. The latter is not very well defined. While the World Meteorological Organization (WMO) recommends 10-minute averages, the U.S. National Hurricane Center and the Joint Typhoon Warning Center both use 1-minute averages. For the classification of individual storm systems and the interpretation of measurements the latter may have considerable consequences. In tropical cyclones near-surface winds are highly turbulent and, compared with 10-minute averages, 1-minute averages usually provide higher wind speeds by about 20% (Kraus and Ebel, 2003). Care is therefore required when data and reports are analyzed and compared, as the averaging period is often not disclosed. Tropical cyclones with wind speeds exceeding $32 \, \mathrm{m \, s}^{-1}$ are further classified according to their damage potential. In the North Atlantic and the East Pacific the Saphir–Simpson scale is used which ranks tropical cyclones with wind speeds exceeding $32 \, \mathrm{m \, s}^{-1}$ into five categories with Category 1 having only small and Category 5 having extreme damage potential. In order for a tropical cyclone to rank as a Category-5 storm, sustained wind speeds must exceed $69 \, \mathrm{m \, s}^{-1}$.

36 **Marine weather phenomena** [Ch. 2

Worldwide there are about 80 tropical cyclones per year, two-thirds of which rank Category 1 or higher and 90% of these storms form within 20°S–20°N (e.g., Gray, 1968; Maue, 2009). Figure 2.8 (see color section) provides an impression of the frequency and the regions influenced by these events. Most tropical cyclones occur over the western North Pacific and may have influence on large parts of the Philippine Islands, the Chinese and Korean coasts, and as far north as Japan. Tropical cyclones also occur in the Bay of Bengal, the Arabian Sea, and the southern Indian Ocean off Madagascar. In the North Atlantic tropical cyclones account for about 12% of the global figure (Shepherd and Knutson, 2007) and may impact the Caribbean Sea, the Gulf of Mexico, and large parts of the U.S. and Mexican coasts. In the eastern North Pacific tropical cyclones may occur off the coast of Mexico sometimes as far west as Hawaii. In the Southern Hemisphere, also the northwest coast of Australia and the South Pacific Ocean from the east coast of Australia to about 140°W are potentially under the impact of tropical cyclones. Regional names for storms that rank Category 1 or higher are typhoons for the western North Pacific, hurricanes for the eastern North Pacific and the North Atlantic, or severe tropical cyclones for the western South Pacific and the Indian Ocean. There are only two ocean basins in the tropics that hardly support any tropical cyclone activity: the South Atlantic and the eastern South Pacific. For the South Atlantic, so far only three tropical cyclones have been reported: the 1991 Angola Cyclone (a tropical storm that formed off Angola), the January 2004 Storm (a cyclone of possibly storm strength) and the 2004 Tropical Cyclone Catarina. Only for the latter do reliable data exist, and it was also the first reported to make landfall as a tropical cyclone (Pezza and Simmonds, 2005). For the eastern South Pacific we are not aware of any tropical cyclone reports. While the Benguela and Peru Currents contribute to lower than normal[6] sea surface temperatures in both basins, it is assumed that the primary reason that the South Atlantic Ocean gets few tropical cyclones is that the tropospheric (near-surface to 200 hPa) vertical wind shear is too strong and that there is typically no intertropical convergence zone (ITCZ) over the ocean (Gray, 1968). Without an ITCZ to provide synoptic vorticity and convergence (i.e., large-scale spin and thunderstorm activity, as well as with strong wind shear), it becomes very difficult to nearly impossible for tropical cyclones to form (Landsea, 2008).

Tropical cyclones are initiated by different atmospheric disturbances with easterly waves (mostly responsible for cyclones over the North Atlantic and the eastern North Pacific), and monsoon troughs (responsible for much of the genesis in the western North Pacific) being the most frequent.[7] They obtain most of their energy from the extraction of latent heat from ocean waters at high temperature and from heat export at the low temperatures of the upper tropical troposphere. Although the exact nature of formation is still an area of extensive research, there are some general large-scale conditions which are believed to favor the formation of tropical

---

[6] Compared with the zonal average.

[7] For details about easterly waves and monsoon troughs see, for instance, Glickman (2000).

cyclones (Shepherd and Knutson, 2007):

1. *A sufficiently warm ocean surface layer with sufficiently large vertical depth.* It is generally concluded that the formation of tropical cyclones requires sea surface temperature above about 26.5°C to 27°C. Moreover, this warm layer must have a thickness of at least about 50 m. As the storms are maintained by the extraction of latent heat from the oceans there will be too small heat fluxes and evaporation rates when temperatures are too low. Similarly, energy supply may be cut when the warm pool is not deep enough as the developing storm will cause Ekman pumping[8] and turbulent mixing of the upper ocean. Precipitation may also have a cooling effect.

2. *A potentially unstable atmosphere to moist convection.* The decrease of temperature with height must be sufficiently strong to allow moist deep convection. When the air is heated at the sea surface it becomes warmer and receives positive buoyancy. Finally the air starts to rise. During its upward motion the air cools by adiabatic processes and loses buoyancy. The ascent is terminated once the rising air is potentially colder than the surrounding environment. However, provided the air has attained sufficient amounts of water vapor near the surface, condensation is initiated and the heat supported by this process provides additional buoyancy enabling further uplift of the rising air.

3. *A moist mid-troposphere.* This will further support the process described above. The ascending air is entraining some air from its surrounding environment. When the latter is relatively humid the process provides additional buoyancy and reinforces convection. When the air entrained is too dry, buoyancy and available convective energy are reduced. The process is most important in the mid-troposphere at about 5,000 m height.

4. *A sufficiently large distance from the equator.* The principle structure of tropical cyclones may be explained by an interplay between pressure gradient, centrifugal force, and Coriolis force (see below). The latter is zero at and increases with distance from the Equator. An intensifying tropical cyclone thus requires a distance of about 500 km from the Equator in order for the Coriolis force to be sufficiently strong to maintain the necessary rotation.

5. *Low values of vertical wind shear.* The wind speed must change only slowly with height. If the change is too strong, the core of the developing system will no longer be vertically aligned and effective deep convection is disrupted. It is assumed that through the troposphere (surface to about 10,000 m height) vertical wind shear should be smaller than about $10\,\mathrm{m\,s^{-1}}$ in order to allow for intensive tropical cyclones to develop. In the literature, the change in wind speed and direction between 850 hPa (about 1.5 km height) and 200 hPa (about 12 km height) is often considered (Vecchi and Soden, 2007).

---

[8] Wind stress will cause a mass transport to the right/left in the Northern/Southern Hemisphere in the upper-ocean layer and is known as Ekman transport. As tropical cyclones have cyclonic wind fields, Ekman transport is directed away from the storm's center causing an upwelling known as Ekman pumping.

38 Marine weather phenomena [Ch. 2

6. *A pre-existing near-surface disturbance with adequate synoptic vorticity (spin) and convergence.* Some atmospheric disturbance such as easterly waves or monsoon troughs are required in the development region to provide the developing storm with initial rotation and convergence. The vertical motion produced by the processes described above is not sufficient and inadequate to initiate rotation. Once sufficiently strong vorticity exists, the developing storm may grow further.

Often large-scale indices are used to summarize the conditions favorable for the development of tropical cyclones and to describe the potential for their formation. An example is the genesis potential (*GP*) index (Emanuel and Nolan, 2004) that includes the effects of vertical wind shear, relative humidity and vorticity at various levels and, via the potential intensity index (Emanuel, 1986), the effects of sea surface temperature and dissipative heating (Camargo *et al.*, 2007). The GP index is defined as

$$GP = |10^5 \eta|^{1.5} \left(\frac{RF}{50}\right)^3 \left(\frac{V_p}{70}\right)^3 (1 + 0.1 V_s)^{-2}, \qquad (2.5)$$

where η is the absolute vorticity at 850 hPa in s^{-1}; RF denotes 700 hPa relative humidity in percent; V_p is potential intensity in $\mathrm{m\,s}^{-1}$ accounting for the effects of SST and dissipative heating; and V_s represents the wind shear between 850 hPa and 200 hPa in $\mathrm{m\,s}^{-1}$. As there are still considerable deficiencies in simulating the long-term behavior of tropical cyclone statistics (such as their frequency and/or intensity per ocean basin) such indices provide an attempt to elaborate on long-term changes despite the existing deficiencies. We will return to this point in Section 5.3.

Tropical cyclones typically consist of an outer region (the borders of which are about 30–50 km and 150 km from the storm's center) where wind speed increases towards the center, an inner region (the eyewall, about 15–30 km from the center) where wind speeds reach their maximum, and the eye where wind speed decreases rapidly towards the center. Here the air is relatively calm and clear skies are prevailing (Figure 2.9, see color section). This structure can generally be explained by the interplay between the different forces involved, the balances of which vary with distance from the storm's center. Large pressure gradients, high wind speeds, and some distance from the Equator indicate the relevant ingredients: pressure gradient, centrifugal force, and Coriolis force. Table 2.1 provides some typical numbers depending on the distance from the storm's center. It can be inferred that in the outer region pressure gradient and Coriolis force are dominant. Eventually this is the reason why the circulation is, contrary to tornados, for example, always cyclonic in tropical cyclones and why they may not be observed too close to the Equator. Approaching the core of the cyclone, the relative contributions from the different forces change and the centrifugal force becomes more and more important. Within the eyewall, the three forces are about in balance. Note that, at the surface, friction is involved additionally and may play a significant role. Approaching the eye, pressure gradients decrease rapidly and become zero at the center. As a consequence wind speed also decreases. Because of turbulent friction with the eyewall (Kraus and Ebel, 2003), wind speed is, however, larger than to be expected solely from the balance of

Table 2.1. Estimates of the magnitude of accelerations (forces) within a synthetic tropical cyclone at $10°$ latitude for given wind speeds and distances from the storm center. Note that friction has been neglected. After Kraus and Ebel (2003).

Distance from storm center (km)	Near-surface wind speed (m s^{-1})	Pressure gradient acceleration (10^{-3} m s^{-2})	Centrifugal acceleration (10^{-3} m s^{-2})	Coriolis acceleration (10^{-3} m s^{-2})	Ratio between centrifugal and Coriolis acceleration
50.0	40.0	33.00	32.00	1.01	31.70
100.0	30.0	9.80	9.00	0.76	11.80
200.0	20.0	2.51	2.00	0.51	3.95
300.0	15.0	1.13	0.75	0.38	1.97
500.0	10.0	0.41	0.20	0.25	0.79

forces. This results in an excess of Coriolis and centrifugal force (which are directed away from the center) over pressure gradient force (directed towards the center) and eventually leads to convergence and upward movement in the eyewall, while downward movement dominates in the eye. Because the air is heated adiabatically when descending, the eye often has clear skies and relatively calm and dry conditions (Figure 2.9).

Some examples of tropical cyclones. The most devastating meteorological phenomena accompanying tropical cyclones are severe winds, heavy precipitation, high storm surges, and over the oceans severe wave heights. Figure 2.10 shows some measurements from a buoy in the Gulf of Mexico for the passage of Hurricane Kate (1985). While at 12:00 UTC sea level pressure was still about 1005 hPa or so, it decreased rapidly afterwards and reached a minimum of less than 960 hPa only 6 hours later. The corresponding pressure tendency was about 8 hPa/hour. For comparison, typical values for extra-tropical cyclones vary around about 1 hPa/hour to 2 hPa/hour. As a result, for Hurricane Kate, peak wind speeds of more than $47\,\mathrm{m\,s}^{-1}$ were observed during its passage over the Gulf of Mexico. Note that wind speeds peaked on both sides of the eye while wind conditions near the center were relatively moderate (less than Beaufort 8). The non-symmetric shape of the wind registration may be associated with the fact that horizontal winds within a tropical cyclone vary in strength depending on the area of the storm in which they occur. The strongest winds are located in the right forward quadrant of the storm (relative to the line along which the storm is moving). The intensification of the winds in this quadrant is caused by the additive effect of winds from the atmospheric flow in which the storm is embedded. Figure 2.11 (see color section) shows a two-dimensional map of the wind field from Hurricane Katrina (2005). Clearly, the relatively moderate wind conditions in the eye can be inferred. Highest wind speeds are obtained in the eyewall. In the outer region wind fields are generally

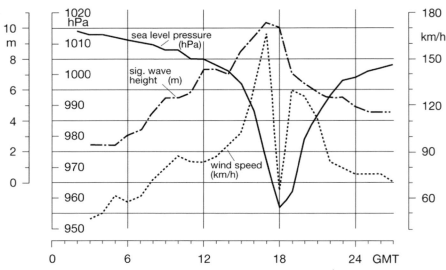

Figure 2.10. Sea level pressure in hPa (solid), wind speed in km h^{-1} (dotted), and significant wave height in m (dashed) as observed from a buoy in the Gulf of Mexico during the passage of Hurricane Kate in 1985. Redrawn after Kraus and Ebel (2003).

cyclonic and wind speed decreases with distance from the storm's center. It can be further inferred that wind fields are highly turbulent and exhibit small-scale, band-like structures as can be obtained schematically from Figure 2.9. For Hurricane Kate (Figure 2.10), ocean wave heights were also recorded. They increased permanently towards the center where they eventually reached peak values of more than 10 m.[9]

Since reliable wind measurements are often lacking, sustained wind speeds are often estimated from core pressures, pressure tendencies, or pressure gradients (e.g., Simpson and Riehl, 1981; Kraus and Ebel, 2003). Table 2.2 shows a list of some of the most intense tropical cyclones in terms of core pressure and related wind speed estimates. It can be inferred that near the center the latter are inconceivably high. The hazard with these numbers is associated with both the direct pressure exerted by the wind and wind gustiness. While the latter may cause resonant oscillations that eventually may result in the failure of structures, wind pressure probably plays the dominant role for wind-related hazards in tropical cyclones. Table 2.3 shows the direct wind pressure associated with wind speeds that are commonly obtained within tropical cyclones. While still depending on the aerodynamical properties of, for instance, a building that is hit by a tropical cyclone and the angle at which the building is being approached, the pressure scales with the square of the wind speed and increases rapidly for Category 4 and 5 cyclones. It can be estimated that, under otherwise similar conditions, the wind pressure exerted by Hurricane Wilma in the

---

[9] Note that these numbers refer to significant wave heights and that individual wave heights may be significantly larger. See Section 2.4 for details.

Sec. 2.3]   Tropical cyclones   41

Table 2.2. Examples of tropical cyclones with unusual deep core pressures compiled from various sources. For comparison Hurricane Katrina is also referenced.

Name	Region	Year	Core pressure (hPa)	Max. sustained wind speed (m s^{-1})
Tip	West Pacific	1979	870	>80
Wilma	North Atlantic	2005	882	>80
Linda	East Pacific	1997	902	>60
⋮	⋮	⋮	⋮	⋮
Katrina	North Atlantic	2005	902	>60

Table 2.3. Estimates of direct wind pressure P according to $P = \dfrac{c_w}{2} A\rho v^2$ where v denotes wind speed, ρ is density, c_w represents the aerodynamical properties of the surface on which the wind acts, and A is the fraction of the surface that is perpendicular to the wind direction. For computation of P it was assumed that $\rho = 1.293\,\mathrm{kg\,m}^{-3}$, $A = 1$, and $c_w = 1$.

Category	Wind speed (m s^{-1})	Wind pressure (Pa)	Ratio	Example
BFT 8	17	192	0.27	Moderate extra-tropical storm, tropical storm
BFT 12	33	704	1.00	Severe extra-tropical cyclone, CAT 1 hurricane
CAT 2	43	1,195	1.70	Hurricane Erin (1995)
CAT 3	50	1,616	2.30	Hurricane Fran (1996)
CAT 4	59	2,250	3.20	Hurricane Charles (2004)
CAT 5	69 80	3,078 4,137	4.37 5.88	Hurricane Andrew (1992) Hurricane Wilma (2005) Typhoon Tip (1979)

vicinity of its eye corresponds roughly to six times that of a severe (Beaufort 12) extra-tropical cyclone.

High wind speeds have also been reported for tropical cyclones with strong pressure gradients or pressure tendencies. Within Tropical Cyclone Tracy (1974, Darwin, Australia) horizontal pressure gradients of 5.5 hPa/km were observed (Kraus and Ebel, 2003). At the airport of Darwin near-surface wind speeds of more

42 Marine weather phenomena [Ch. 2

Table 2.4. Examples of tropical cyclones with unusually high death toll or damage compiled from various sources. Damage is provided in terms of USD as of the year indicated in parentheses. Data compiled from various sources.

Name	Region	Year	Damage (10^9 USD)	Fatalities
Katrina	North Atlantic	2005	81.2 (2005)	~1,900
Andrew	North Atlantic	1992	44.9 (2005)	65
Great Hurricane	North Atlantic	1780	—	~22,000
Mitch	North Atlantic	1998	6.0 (2006)	11,000–18,000
Galveston	North Atlantic	1900	—	8,000–12,000
Rusa	Western North Pacific	2002	6.0 (2002)	113
Bilis	Western North Pacific	2006	4.4 (2006)	672
Thelma	Western North Pacific	1991	0.019 (1991)	6,000
Nargis	North Indian Ocean	2008	10 (2008)	146,000

than $60\,\mathrm{m\,s^{-1}}$ were measured before the instrument was destroyed.[10] Strong pressure tendencies have been reported for Typhoon Irma (1971, 78 hPa within 12 hours) and Hurricane Kate (1985, more than 40 hPa within 6 hours, Figure 2.10). Apart from wind, precipitation and storm surges are additional hazards associated with landfalling tropical cyclones. For example, extreme precipitation was reported for Tropical Cyclones Denise (1966) and Hyacinthe (1980) both affecting the island of Réunion in the Indian Ocean. For Denise 1,144 mm rainfall within 12 hours and 1,825 mm within 24 hours were reported. Figures for Hyacinthe were even higher with 5,876 mm within 10 hours and 3,240 mm within 72 hours (Kraus and Ebel, 2003). Examples of severe storm surges associated with tropical cyclones are provided in Section 2.5.

Table 2.4 lists some of the most severe tropical cyclones in terms of damage and death toll. Examples of damage associated with the Galveston Hurricane (1900) and Hurricane Katrina (2005) are provided in Figure 2.12. Noticeably it is not always the most intense or severe tropical cyclones that caused the highest damage or fatalities. For example, Hurricane Katrina represents one of the most devastating tropical cyclones impacting the U.S. Gulf Coast. However, in terms of core pressure, wind speed, and behavior, Hurricane Katrina was less exceptional. The storm was first identified as a tropical depression over the southeastern Bahamas and it intensified to a Category-1 hurricane shortly before striking Miami (McTaggart-Cowan *et al.*, 2007). Crossing the Florida peninsula the storm showed little weakening which

---

[10] *http://en.wikipedia.org/wiki/Cyclone_Tracy*, last accessed March 25, 2008.

Sec. 2.3] Tropical cyclones 43

Figure 2.12. Damage caused by wind and storm surge during the Galveston Hurricane in Galveston, Texas in September 1900 (top) and flooding and levee repair in New Orleans, Louisiana after the passage of Hurricane Katrina in September 2005. From *http://www.photolib.noaa.gov*, reproduced with permission from NOAA.

was shown to be *not* unusual by McTaggart-Cowan *et al.* (2007) who compared the behavior of Katrina with those of comparable hurricanes in the area. In particular, McTaggart-Cowan *et al.* (2007) concluded that the minor zonal extent of the peninsula, the lack of orographic features, and the available moisture in the swamp-like Everglades hamper significant disruption of the storm when passing over Florida. Eventually Hurricane Katrina rapidly intensified over the warm waters of the Gulf of Mexico before weakening and making a pair of destructive landfalls in Louisiana, demonstrating the fragility of the coastal defense infrastructure bordering the Gulf of Mexico (McTaggart-Cowan *et al.*, 2007). In consequence, Hurricane Katrina represented a severe, but not an extremely severe and unusual hurricane, whose impacts had been enormously magnified by a lack in both adequate preparation and response (van Heerden and Bryan, 2006).

2.4 WIND-GENERATED WAVES

2.4.1 Introduction

Wind waves are generated by the action of the wind on the sea surface. If a wind starts blowing over an initially calm sea surface it first establishes a very shallow surface current whose velocity is proportional to the wind speed (van Dorn, 1994). At some point turbulent fluctuations develop within the air above the sea surface. They disturb the water surface and a characteristic pattern of capillary waves appears. These waves are so small that gravity is not important and surface tension represents the major restoring force. When the wind speed raises further, the wavelength, period, and height of capillary ripples increase. The ripples are still too small to significantly disturb the air flow above. At some point, the wind speed is sufficiently strong to generate waves that are high enough to perturb the air flow above them. This further increases the turbulence of the air flow and turbulent pressure fluctuations tend to further increase the wave height. This process is directionally selective and strongest for wave components traveling nearly into the direction of the wind (van Dorn, 1994).

When the waves get higher, gravity is becoming more and more important and replaces surface tension as the major restoring force. Capillary waves have wavelengths shorter than about 2 cm while surface gravity waves, for which gravity is the dominant restoring force, have wavelengths larger than approximately 10 cm. In the intermediate length range both restoring forces, surface tension and gravity, are about equally important. Because of the different physics of capillary and gravity waves, the speed of capillary waves decreases with increasing wavelength while the speed of gravity waves increases with increasing wavelength. As a result, there exists a smallest possible phase speed for wind-generated waves, which is about $0.23 \, \mathrm{m \, s^{-1}}$ for waves with wavelengths of about 1.73 cm. It can be inferred that the initial speed of a capillary wave is too small to follow the wind speed. As a consequence, waves initially travel at an angle of about $70°–80°$ from the wind direction (van Dorn, 1994). When the waves grow and their speed increases this angle progressively decreases.

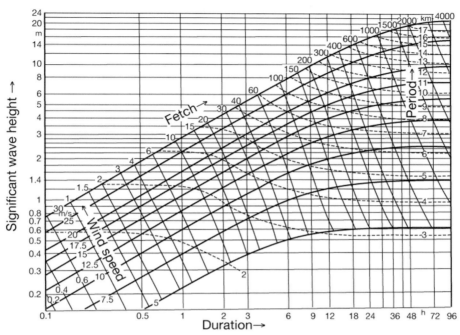

Figure 2.13. Dependence of significant wave height in meters and wave period in seconds on near-surface wind speed in m s^{-1}, duration in hours, and fetch in kilometers. Redrawn after WMO (1998).

For deep water, waves at the sea surface are generally determined by three factors: wind speed, duration, and fetch. Here *duration* denotes the period of time over which the wind is blowing and *fetch* refers to how much sea room there is available for the wind to build up the waves. The relation between wind speed, duration, fetch, and wave height is illustrated in Figure 2.13. Thus a moderate extra-tropical storm with wind speeds of about 17.5 m s^{-1} (Beaufort 8) would generate waves of about 3.5 m significant wave height[11] after 6 hours and about 6 m after 18–24 hours provided the fetch is sufficiently large. When both fetch and duration are sufficiently large, a sea state generally known as *fully developed sea* is eventually reached. In this case, the phase speed of the waves has grown to a value rather similar to that of the near-surface wind speed. Hence the energy input into the waves by the wind field is largely balanced by dissipation processes such as white capping or wave breaking, and the waves cannot grow any further. For small wind speeds less time and fetch are needed to obtain fully developed seas. According to Figure 2.13, a wind speed of 5 m s^{-1} would result in fully developed seas within 18 hours provided

---

[11] Note that the diagram refers to *significant* wave height (see Section 2.4.2) and that individual waves may be considerably higher. As an approximation the significant wave height represents the average conditions of the top third of waves observed over a defined period.

46   **Marine weather phenomena** [Ch. 2

the fetch is about 150 km or more. For wind speeds of $20 \, \text{m s}^{-1}$ about 72 hours and 2,000 km fetch are required. As a rule of thumb, the significant wave height for fully developed seas H_{fds} can be obtained from

$$H_{\text{fds}}(\text{m}) = 2.45 \left(\frac{u_{10}(\text{m s}^{-1})}{10} \right)^2 \tag{2.6}$$

where u_{10} denotes the wind speed at 10 m height. Thus, for wind speeds of about $10 \, \text{m s}^{-1}$ H_{fds} is about 2.45 m, while for $20 \, \text{m s}^{-1}$ it is about 10 m. Further, as a rule of thumb, the corresponding wave period T_{fds} can be obtained from

$$H_{\text{fds}}(\text{m}) = \frac{1}{20} T_{\text{fds}}^2(\text{s}). \tag{2.7}$$

Usually for high-wind speeds fetch and duration are seldom unlimited and observed wave heights are thus considerably smaller. Generally, the fetch depends on the sea area and the direction from which the wind is blowing. For instance, for the Strait of Gibraltar the fetch is about 930 km for winds coming from the east to the northeast while it is only a few kilometers for winds coming directly from the north or from the south. For the northern parts of the Adriatic Sea and southeast winds the fetch is about 750 km, while it is about 830 km for the German Bight (southeastern North Sea) for northwest winds. Duration is intimately connected with the time scales of the meteorological phenomena causing high wind speeds. It may range from several hours for local high wind speed phenomena such as mistral or etesian winds over the northwestern Mediterranean or the Aegean Seas, respectively, to several days for a series of extra-tropical cyclones in the storm tracks of both hemispheres. Situations in which fetch and/or duration are limited are referred to as *fetch-limited* and/or *duration-limited* growth.

As the spatial extent and the duration of a storm are limited, so is the area where the action of wind causes the waves to develop. Waves that are still forced by the local wind field and that, as a consequence, are still growing are referred to as *wind sea*. Wind sea generally travels parallel to the wind direction. When wind direction and/or wind speed change, a new wind sea system develops that corresponds to the new wind field. The old waves still exist, but are decoupled from the wind forcing. Such waves are referred to as *swell*. Swell may propagate over long distances with only little damping. As a result there are often different wave systems existing simultaneously at a location. These systems are usually characterized by different periods, directions, and wave heights. They are often represented in the form of *wave spectra*; that is, wave energy as a function of wave period and direction.[12] Figure 2.14 shows an example of such a wave spectrum with two different wave systems. First, a wind sea system propagating approximately towards the northeast can be inferred. The system is driven by a local wind field with directions from the southwest. The wind sea system has a peak frequency of about 0.3 Hz, which corresponds to a period of about 3 s. The wavelength is in the order of 15 m. Second, a swell system with maximum energy at

---

[12] See Section 3.4.1 for a more detailed discussion of wave spectra.

Sec. 2.4] Wind-generated waves 47

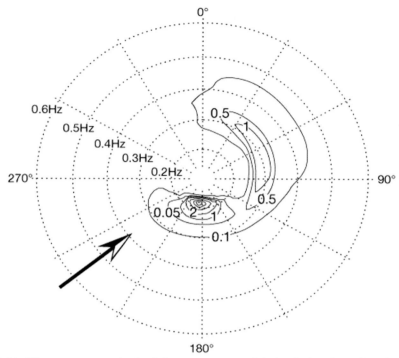

Figure 2.14. Wave spectrum obtained from a wave model simulation for a location in the central North Sea (55.5°N, 0.5°E). The total significant wave height at the time of the simulation (12:00 UTC, February 15, 1962) is about 3.15 m; the mean wave direction is about 161°. Wind is coming from 232° with a speed of about 12 m s^{-1}. In the plot, distance from the center denotes frequency; direction is indicated at the outermost circle.

about 0.15 Hz to 0.2 Hz is inferred. Typical wavelengths in this system are about 60 m, and the system moves towards the south. Both systems travel at an angle of about 120°, a situation which is usually referred to as a *cross sea*. Cross seas generally lead to a "confused" sea state. As ships in heavy seas usually travel against the waves cross seas may represent a particular threat for navigation.

Figure 2.15 (see color section) illustrates how swell may propagate over long distances. Shown is an example in which waves produced by a low-pressure system over the northwest Pacific traveled over about 1,000 km to 1,500 km until they reached North Shore, Oahu, Hawaii. The strong wind fields over the northwest Pacific forced a wave system with wave heights well above 12 m. The system propagated towards Hawaii and eventually reached Hawaiian deep waters with wave heights of about 6 m. When the swell system moved further into the shallower coastal waters, the waves began to slow down and the wave heights increased because of the divergence in wave group velocity (see Section 3.4.1), a process referred to as *shoaling*. As a result, surf heights of more than 12 m were reported from North Shore, Oahu. An impression of such waves is given in Figure 2.16.

48 Marine weather phenomena   [Ch. 2

Figure 2.16. High surf conditions along North Shore, Oahu (Hawaii) on November 16, 2004.

2.4.2 Long and short-term variations of the sea state

When dealing with the variability of wind-generated waves at the sea surface it is important to distinguish between two fundamentally different concepts: the variations *within* a given sea state (short-term variations) and the variations *of* the sea state itself (long-term variations). For an observer at a fixed location the sea surface elevation varies considerably within seconds caused by the passage of the crests and troughs of individual waves. On the other hand, even under these circumstances there exist some properties that describe the prevailing conditions relatively well and which remain almost constant over some period of time. An example is the *average* height of the waves generated within a particular extra-tropical storm. Hence, such parameters are generally used to describe, in a statistical sense, the actual conditions of a wave field. They are also used to describe *long-term* changes such as the variations from day to day, month to month, or year to year, or in other words, the variations *between different* sea states. In the following we refer to such changes as *long-term* variations.

For any given set of such parameters describing the actual conditions or the sea state in a statistical sense, there still remains considerable variability due to the variations of the sea surface caused by individual waves; that is, variations that occur *within* a given sea state. We will refer to such fluctuations as *short-term* variations. Figure 2.17 illustrates the problem. The upper panel shows a record of the measured sea surface elevations over a period of 20 minutes. The up and down movements in the time series are caused by the passage of individual waves and it can be inferred

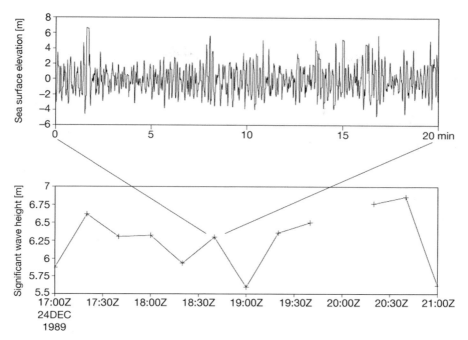

Figure 2.17. Time series of sea surface elevation measured over a 20-minute time interval (top) and corresponding significant wave height derived from 20-minute measurements over a time period of 4 hours (bottom). The cross at 18:40 UTC in the lower panel represents the significant wave height derived from the 20-minute time series shown in the upper panel.

that the height of these individual waves varies considerably. When averaging over the top 33% of the individual waves is performed a characteristic number referred to as *significant wave height*[13] is derived that is representative for the actual conditions. From the lower panel it can be obtained that this number is about 6.25 m for the time interval 18:30–19:50 UTC shown in the upper panel. The lower panel also shows the variations of the significant wave height itself when a longer period of 4 hours is considered. While still relatively constant within this 4-hour period it can be envisioned that significant wave heights vary on longer time scales such as those associated with the passage of low-pressure systems, between seasons, and on longer time scales such as years and decades.

Distribution of waves within a given sea state: the Rayleigh distribution. For many practical applications short-term variations of the sea state are important and critical. Generally, the design of offshore structures or vessels is based on the long-term statistics of parameters describing the sea state; in particular, on which conditions the structure or the ship is likely to encounter during its lifetime. On the other hand,

---

[13] See Section 3.4.3 for details.

50   Marine weather phenomena [Ch. 2

it is not the *average* conditions that cause a structure to fail but the most extreme individual waves. A typical questions in marine design is therefore: *What is the highest wave a structure is likely to encounter during its lifetime?* Or, in other words: *Is there a distribution that describes the probability of individual wave heights for a given sea state?*

An approximation of such a distribution can indeed be inferred. By considering the height h of individual waves within a given sea state as a random variable, where the sea state can be characterized by the average wave height $H_{1/3}$ of the top third of the individual waves, Longuet-Higgins (1952) showed that a probability distribution for h can be approximated by

$$F(h < h_p) = p$$

$$= 1 - \exp\left[-2\left(\frac{h_p}{H_{1/3}}\right)^2\right]. \tag{2.8}$$

Equation (2.8) is generally known as the *Rayleigh distribution*. The parameter $H_{1/3}$ that describes the sea state is often referred to as the *significant wave height*, a concept which was originally introduced by Sverdrup and Munk (1947). Roughly, the significant wave height corresponds to the height that an experienced observer at sea would report as the *typical* wave height for the prevailing conditions (sea state).

The wave height h_p in (2.8) is called the *p-percentile* or the *p-quantile*. For a given significant wave height, the *p*-percentile is exceeded by only $(1 - p)$ percent of all individual waves, while the remaining p percent are smaller than h_p. The probability $(1 - p)$ is therefore also called the *exceedance probability*

$$1 - p = 1 - F(h < h_p) = F(h > h_p). \tag{2.9}$$

Table 2.5 shows the exceedance probabilities for different wave heights derived from the Rayleigh distribution. It can be inferred that about 14% of individual waves will be higher than the significant wave height, a fact that also follows directly from the definition of $H_{1/3}$. Further it can be inferred that only 1 out of 100 individual waves is likely to exceed the significant wave height by 50% and that the exceedance probability for observing a wave with twice the significant wave height is only about 3.4×10^{-4}. Therefore, the maximum wave height likely to be observed within a particular sea state is often estimated by, as a rule of thumb, twice the significant wave height.

Table 2.5. Ratio of individual wave height h_p and significant wave height $H_{1/3}$ and the related exceedance probabilities $F(h > h_p)$ determined from the Rayleigh distribution (2.8).

$\dfrac{h_p}{H_{1/3}}$	1.0	1.5	2.0	2.2
$F(h > h_p)$	0.14	0.01	3.4×10^{-4}	1.0×10^{-4}

Given the Rayleigh distribution the question remains what would be the highest wave an observer is likely to encounter for a given sea state lasting for a given period. The answer is directly related to the number of waves passing the observer within the given period of time. Suppose the sea state persists for a time period T in which the observer encounters N individual waves. Suppose further that the amplitudes of these waves are randomly distributed. Then the probability distribution (2.8) for the largest wave h_{max} to be expected reads

$$F(h < h_{max}) = 1 - \frac{1}{N}$$

$$= 1 - \exp\left[-2\left(\frac{h_{max}}{H_{1/3}}\right)^2\right], \qquad (2.10)$$

which by rearrangement yields

$$h_{max} = H_{1/3}\sqrt{\frac{\ln N}{2}}. \qquad (2.11)$$

Let us a assume that a storm with high wind speeds lasts for about 6 hours and produces a sea state with typical wave periods of about 18 s. Then the observer will notice approximately 1,200 individual waves during the course of the storm. According to (2.11), the highest wave the observer will encounter is then about 1.88 times the significant wave height. For a wind speed of about $20\,\mathrm{m\,s}^{-1}$ and fully developed seas, producing a significant wave height of about 10 m (Figure 2.13), the highest wave likely to be observed within 6 hours therefore is about 18.8 m.

Note that the time it takes, on average, to observe an individual wave with a height two times or more the significant wave height depends on the sea state. By rearranging (2.11) with respect to N it can be obtained that about 1 out of 2,980 individual waves will be larger than twice the significant wave height. As a rule of thumb wave height and wave period are coupled via (2.7). From (2.7) it can be inferred that higher waves are linked with longer periods and *vice versa*. For small waves, the time it needs to observe 2,980 individual waves is therefore smaller than for higher waves. For example, for fully developed seas a wind speed of $10\,\mathrm{m\,s}^{-1}$ results in a significant wave height of about 2.45 m and a wave period of about 7 s according to (2.6), (2.7). The time it takes to encounter 2,980 waves is therefore about 6 hours. For a wind speed of $20\,\mathrm{m\,s}^{-1}$, the significant wave height and period for a fully developed sea state are 9.8 m and 14 s, respectively. The time it takes to pass 2,980 waves (or to observe, on average, a wave of more than twice the significant wave height) is therefore almost 12 hours or twice as large.

2.4.3 Freak or rogue waves

From time to time reports are made about ships or offshore platforms that received damage from unexpectedly high and/or steep individual waves. Such waves are usually referred to as *freak* or *rogue* waves. An example is shown in Figure 2.18. It shows the oil freighter *Esso Languedoc* off the coast of Durban (South Africa),

Figure 2.18. Freak wave approaching the oil freighter *Esso Languedoc* in 1980. Credits: Philippe Lijour/ESA.

when an extremely high wave approached the vessel from behind and was breaking on the foredeck of the ship. At the time of the incident the significant wave height was reported to be between about 5 m to 10 m. The crest height[14] of the freak wave can be vaguely inferred from the mast on the starboard side of the foredeck which has a height of about 25 m above mean sea level. In this case the incident caused only minor damages. In 1974 the Norwegian freighter *Wilstar* (Figure 2.19) was less fortunate. According to the crew a rogue wave off the South African coast ripped parts of the bow. The damage was caused by a combination of pitch motion and a steep incoming wave. Some more examples of freak wave occurrences are compiled in Table 2.6 and further examples can be found, for instance, in Kjeldsen (1997) or Didenkulova *et al.* (2006).

While the existence of rogue waves is now generally accepted, a broad consensus regarding their definition is still lacking. In many studies individual waves are considered to be rogue waves when their height exceeds about twice the significant wave height. Other studies use crest height thresholds of 1.25 (e.g., Haver and Andersen, 2000a) or similar. According to the Rayleigh distribution (2.8) such events are highly

---

[14] Crest height refers to the height of the wave above the still water line. Wave height, on the other hand, is measured as the distance between the crest and the preceding or following trough.

Figure 2.19. Impact of a freak wave on the Norwegian freighter *Wilstar* in 1974. Credits: ESA/DLR.

unlikely.[15] When more sophisticated models of the sea surface[16] are adopted, the chances of observing freak waves increase, but still remain small.

The question remains on whether freak waves may be considered as rare realizations of a typical wave population or whether it is appropriate to assume the existence of a separate freak wave population. When waves are considered to be realizations of a random process, there always remains some chance of observing extreme (freak) waves. An intuitive example is provided by Haver and Andersen (2000a) who supposed a 10-year storm[17] that blows over an ocean area large enough that the waves in 100 sub-areas can be assumed to be statistically independent. In this case, it is very likely that a wave close to a (local) 1,000-year event can be observed in one of the sub-areas. In this example, the observed extreme (freak) wave represents a rare realization of a typical wave population. An alternative approach is to assume the existence of a separate freak wave population. The latter was suggested, for instance, by Haver and Andersen (2000b). Here freak waves represent typical realizations of a rare population, a circumstance with considerable consequences: being members of a separate population, freak waves may occur more often than could be assumed solely from the probability of rare events for typical wave populations. Although the existence of such a separate freak wave population is far from being proved it receives some support from the relatively large amount of reported ship damage and number of incidents that are attributed to the impact of freak waves.

---

[15] Note that the time period it takes for observing an individual wave height exceeding the significant wave height by a certain factor depends on the significant wave itself (see Section 2.4.2).
[16] Such as second-order probabilistic models (e.g., Haver and Andersen, 2000a).
[17] That is, a storm that occurs on average only once every 10 years.

54   **Marine weather phenomena**   [Ch. 2

Table 2.6. Some examples of freak wave incidents compiled from Haver (2000) and Didenkulova *et al.* (2006).

Location	Description
Off eastern South African coast	The ship-master of the *Edinburgh Castle* described an event during which the vessel was hit by a freak wave in 1964
Southwest of Durban (South Africa)	*Cruiser Birmingham* was hit by a freak wave in 1944
Off Newfoundland	In 1995 the *Queen Elisabeth II* headed into a major storm, Luis. The maximum wave height from the ship log was close to 30 m. The shipmaster described an episode where the crew were looking at a wall of water for a couple of minutes before it hit the ship
Off the east coast of Greenland	In 1943 the liner *Queen Elisabeth* hit a trough preceding a giant wave. The wave impact shattered the bridge window about 90 feet above the normal water line
North Sea	The Statoil-operated Draupner platform was hit by a freak wave in January 1995
Mediterranean Sea, 60 miles off Menorca	In 2005 the cruise vessel *Grand Voyager* was hit by an about 14 m high individual wave within a sea state of only 2.4 m significant wave height. The wave caused some damage and injuries onboard
Atlantic Ocean	Between the Bahamas and New York the cruise vessel *Norwegian Dawn* encountered a more than 21 m high freak wave in 2005. Maximum significant wave height was reported to be about 4 m

So far most reports on the existence of freak waves are based on anecdotal evidence or photographs of damage that is attributed to the impact of freak waves. Only a limited number of measurements does exist. An example is the so-called *New Year Wave*, a freak wave that hit the Statoil-operated Draupner platform in the northern North Sea on January 1, 1995. The time series obtained from a down-looking laser device is shown in Figure 2.20. During a 20-minute period shown in the time series individual waves with typical heights mostly between about 10 m to 12 m were observed. The significant wave height during that period was reported to be about 12 m (Haver and Andersen, 2000a). In the first quarter of the period shown an

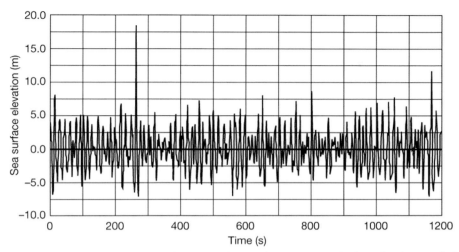

Figure 2.20. Time series of sea surface elevation at the Draupner platform in the North Sea. Courtesy of Sverre Haver, Statoil.

exceptionally high and steep wave passed the laser device. With a height close to 26 m it measured more than twice the significant wave height and the crest height was reported to be about 18.5 m. The latter implies that the wave was accompanied by a rather shallow trough and it may well have appeared to an observer as a "wall of water", a description that is often provided by observers having experienced a rogue wave event. Some more examples of freak wave measurements can be found, for instance, in Sand *et al.* (1990), Kjeldsen (1997), and Yasuada *et al.* (1998).

While interpretation from time series may provide some insight into the time behavior of the phenomenon, the question about its regional distribution remains. So far, approaches other than to study the regional distribution of ship accidents in combination with wave hindcasts (e.g., Monbaliu and Toffoli, 2003) are limited. Only recently have algorithms been developed that are supposed to measure individual wave heights from spaceborne synthetic aperture radars (SAR) mounted aboard ERS satellites (Rosenthal *et al.*, 2003). An example of such a measurement is provided in Figure 2.21. It shows a cross-section of the surface elevation over a range of 5 km taken from ERS/SAR data over the Southern Ocean. From the cross-section irregularly varying waves typically of about 10 m height can be inferred. An extreme outlier is the individual wave in the middle of the cross-section. It has a height of 29.8 m which is about 2.9 times the significant wave height reported for the time the satellite data were taken. According to current definitions, the wave thus qualifies as a freak or rogue wave. Unfortunately, only a limited number of SAR data have been processed so far. While satellite data may provide a tool to gain further insight into the regional distribution of freak waves, further data analysis is needed before reliable statements can be made.

Freak waves constitute an enormous concentration of wave energy. The mechanisms leading to such concentrations are still under discussion and are subject

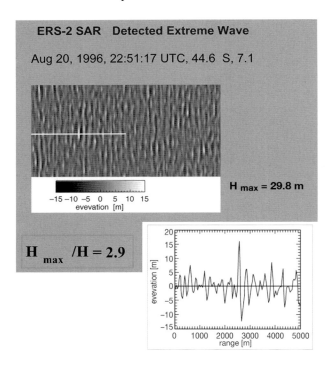

Figure 2.21. Freak wave detected from ERS-2 SAR in August 1996 in the Southern Ocean. The upper panel shows the picture derived from the satellite data. The cross-section shown in the lower panel is indicated by the white solid line. Courtesy of Susanne Lehner, German Aerospace Center.

of intensive research. So far, in principle three different approaches exist, all of which have their advantages and limitations:

1. *The random concentration of wave energy in time and space* mainly as a consequence of the dependence of phase speed on wavelength. Here freak waves may develop as a result of a superposition of different wave components traveling at different speeds. The limitation of this approach is the required delicate balance of initial conditions which may well be simulated in wave tanks, but which are hardly realized spontaneously in the open ocean, and the relatively short life time of the freak waves which is not in agreement with observational evidence.
2. *The wave energy concentration caused by ocean currents.* In this case currents acting over long distances deflect the waves such that local focusing or defocusing of wave energy occurs. In order for this mechanism to be effective all wave components need similar directions when entering the current. Such a situation may indeed be encountered at a few places such as the Agulhas Current. However, freak waves have been observed also in areas that lack strong ocean currents.
3. *Energy focusing due to non-linear effects.* Such mechanisms have been proposed by various authors. In the simplest of these approaches non-linear effects account for a redistribution of wave energy such that a small part of a wave train is growing at the expense of its surroundings. For a summary see Haver and Andersen (2000b).

2.5 TIDES, STORM SURGES, AND MEAN SEA LEVEL

Variations in the height of the sea surface are caused by various factors that extend over a wide range of space and time scales. When we ignore variations caused by wind waves,[18] the observed sea surface height h_t can, for many practical purposes, be regarded as a superposition of tides, surges, and changes in mean sea level

$$h_t = t_t + s_t + l_t, \tag{2.12}$$

where the index t denotes time; and t_t represents the tidal variations that are caused by astronomical forces such as the gravitational forces exerted by the Moon and the Sun. These forces cause complicated patterns in ocean response, but are in general regular and predictable; s_t denotes the surges which are caused by meteorological factors such as the wind acting on the sea surface and pushing waters towards or away from the coast. Compared with tides, surges are less predictable and regular. They alter, sometimes significantly, the tidal pattern that dominates the variations in sea surface height at most places and on short time scales. On long time scales changes in mean sea level l_t are particularly important. They refer to changes in sea surface height that are caused by slowly varying factors such as geological changes (e.g., vertical land movements), melting of grounded ice, or thermal expansion of sea water caused by long-term changes in ocean temperatures. In the following we will discuss the three terms on the right-hand side of (2.12) in more detail.

2.5.1 Storm surges

The word *surge* generally refers to a rise or drop of the sea surface caused primarily by winds and secondarily by the action of atmospheric pressure on the sea surface. Significant surge heights are generated when the wind pushes large water masses towards or away from the coast. The effect should not be mixed up with sea surface height variations caused by wind waves, which have much shorter wavelengths, space and time scales (see Section 2.4). More generally and neglecting for slow changes in mean sea level and for tide–surge interactions (e.g., Horsburgh and Wilson, 2007), surges are also referred to as *non-tidal* or *meteorological* residuals

$$s_t = h_t - t_t. \tag{2.13}$$

Depending on whether the water is transported towards or away from a location, surges may enhance or reduce the local sea surface height. In the first case they are referred to as positive surges, in the second case as negative surges. While positive surges are generally more spectacular, especially when they are associated with the appearance of major storm systems (*storm surges*), negative surges may, in some cases significantly, reduce local water depths and may represent a potential hazard for shipping. Table 2.7 provides a list of some major storm surge events and their estimated impacts. It can be inferred that the highest surges and the most destructive

---

[18] These are variations in the sea state, which are treated in detail in Section 2.4.

58   **Marine weather phenomena**   [Ch. 2

Table 2.7. Estimated impact of some major storm surge events compiled from various sources.

Date	Region	Name or type of storm system	Max. surge level (m)	Approx. number of fatalities
September 1900	Galveston, Texas	Galveston Hurricane	4.6	6,000–12,000
February 1953	Southern North Sea (U.K., Netherlands)	Unnamed extra-tropical cyclone	3.0	2,000
February 1962	Southern North Sea (Hamburg, Germany)	Extra-tropical Cyclone Vincinette	3.5	325
November 1970	Bangladesh	Bhola Tropical Cyclone	6.0–10.0	300,000–500,000
April 1991	Bangladesh	Bangladesh Tropical Cyclone	5.0–8.0	150,000
August 2005	Alabama, Mississipi, Louisiana (U.S.A.)	Hurricane Katrina	8.2	1,900

events are often associated with tropical cyclones. Although the loss of life caused by tropical cyclones is not entirely caused by the storm surge, the latter is responsible for the majority of fatalities in tropical cyclones worldwide. For example, for the Bhola Tropical Cyclone that hit Bangladesh in November 1970 the official estimate was put at about 300,000–500,000 fatalities, making the storm the deadliest tropical cyclone on record. It was estimated that about 46,000 fishermen operating in area at the time of the storm lost their lives. More than 9,000 fishing boats were destroyed and the damage to property and crops was enormous. The maximum recorded wind speed was about $60\,\mathrm{m\,s^{-1}}$ and the maximum storm surge height was estimated to be in the order of about 10.0 m.[19] A comparable disaster in the area was caused by the tropical cyclone that hit Bangladesh in April 1991 resulting in about 150,000 fatalities. In the United States, the storm surges caused by the Galveston Hurricane and by Hurricane Katrina so far represent two of the most destructive examples.

The spatial and time scales of storm surges vary considerably (e.g., Gönnert *et al.*, 2001). Tropical cyclones are associated with small and intense surges that have typical time scales of a few hours and spatial scales on the order of about 50 km. Figure 2.22 schematically illustrates the situation for a Northern Hemisphere tropical storm. The peak of the storm surge occurs along the forward right quadrant of the landfalling storm. Here onshore winds push the water masses towards the coast. To the left of the storm's center surge heights are greatly reduced or even negative as a result of offshore winds. In reality the situation may become more complex.[20] Extra-tropical storm surges, in contrast, typically have much larger spatial dimensions of up to several hundreds of kilometers, and longer time scales on the order of one day.

---

[19] Source: *http://en.wikipedia.org/wiki/1970_Bhola_cyclone*, last accessed March 27, 2008.
[20] See example on Hurricane Dennis later in this section.

Sec. 2.5] Tides, storm surges, and mean sea level 59

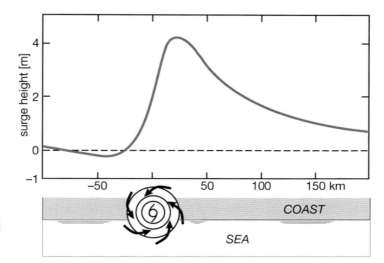

Figure 2.22. Schematic diagram of the distribution of storm surge heights along a coastline during landfall of a tropical cyclone. Redrawn after Liu (2004).

They represent relatively slow-moving phenomena that may affect large areas over periods of up to several days. They may also propagate away from the storm, enter the continental shelf, and propagate as long waves along the coast. In the latter case they are usually referred to as *external surges*.[21]

The contribution of surges to local variations in sea surface height can be assessed from the statistics of long time series. An obvious measure for the magnitude of weather-related effects on sea surface height variations is the standard deviation obtained from hourly surge data. Figure 2.23 shows an example for Cuxhaven located in the German Bight (southern North Sea) at the mouth of the River Elbe. The standard deviation of observed surges within the period 1958–2002 is about 36 cm. For comparison, the standard deviation of observed surge heights varies between a few centimeters for ocean islands surrounded by deep water (e.g., 6 cm for Honolulu, Hawaii) to tens of centimeters for shallow waters subjected to frequent stormy weather conditions (e.g., 20 cm for Southampton, U.K.) (Pugh, 2004). Largest values are often observed in estuaries or bays, such as shown here for Cuxhaven (36 cm) or Buenos Aires (49 cm) (Pugh, 2004). Further inspection of Figure 2.23 reveals that the observed distribution has extended tails for both positive and negative surges with a tendency for large positive values to occur more frequently. The largest positive and negative surges observed between 1958 and 2002 are 3.50 m and −2.25 m, respectively.

Seasonal variations in surge heights may be assessed, for instance, from changes in monthly surge levels that are exceeded for only 1% of the time (99-percentiles). For Cuxhaven these levels are shown in Figure 2.24. Here the seasonal variations in surge statistics basically reflect the seasonal cycle of weather patterns. The most severe surges tend to occur in the storm season from November to February

---

[21] See example on the North Sea storm surge in 1962 later in this section.

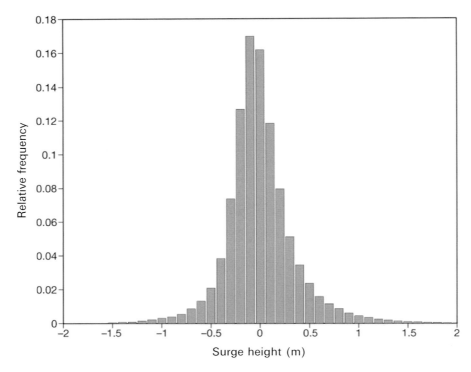

Figure 2.23. Relative frequency of observed storm surges in Cuxhaven obtained from hourly data 1958–2002. The analysis was performed using a 10 cm bandwidth. The standard deviation of the observed surge is 36 cm, the largest positive and negative surges were 3.50 m and −2.25 m, respectively.

while the chance of severe storm surges is considerably reduced during the summer months.

Inverse barometric effect. When atmospheric pressure at the sea surfaces rises, the height of the sea surface is depressed and *vice versa*. This effect is known as the *inverse barometric effect*. Once the ocean has reached an equilibrium in response to atmospheric pressure forcing, the following equation holds:

$$p_a + \rho g h = \text{const.} \qquad (2.14)$$

where p_a denotes atmospheric pressure at the sea surface (sea level pressure); ρ is the density of seawater; g is gravitational acceleration; and h represents the height of the sea surface. Any change in sea level pressure Δp_a will thus lead to a corresponding change in sea surface height

$$\Delta h = -\frac{\Delta p_a}{\rho g}. \qquad (2.15)$$

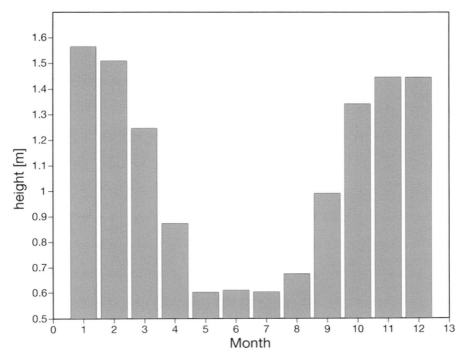

Figure 2.24. Monthly 99-percentile surge heights obtained from hourly data at Cuxhaven 1958–2002.

When average values of $\rho = 1{,}026\,\text{kg}\,\text{m}^{-3}$ and $g = 9.81\,\text{m}\,\text{s}^{-2}$ are taken for density and gravitational acceleration, then, as a rule of thumb, a change in atmospheric pressure of about 1 hPa will lead to a 1 cm change in sea surface height caused by the inverse barometric effect. For an extra-tropical cyclone with a core pressure of 950 hPa the inverse barometric effect will thus result in a sea level rise of about 63 cm relative to the average atmospheric sea level pressure of 1013 hPa. On January 10, 1993 an extremely intense cyclone with a core pressure of 913 hPa was observed north of the United Kingdom (Kraus and Ebel, 2003). In this extreme case, the inverse barometric effect may, in theory, have caused a sea level rise of up to 1 meter in the vicinity of the center of the low-pressure system. Usually, however, sea level pressure varies between about 980 hPa and 1,030 hPa corresponding to inverse barometric effects between about +33 cm and −17 cm relative to the height of the sea surface at 1013 hPa. Therefore, the inverse barometric effect is usually of secondary importance compared with the direct response of sea surface height to wind forcing. It should be noted, however, that the inverse barometric effect may be enhanced by a type of resonance for a moving atmospheric pressure system, especially when the latter moves over a continental shelf with a speed close to that of a shallow-water wave (Pugh, 2004).

62 **Marine weather phenomena** [Ch. 2

Response of sea surface height to wind forcing. When the wind blows over the ocean, the drag of the sea surface acts in a way to change sea surface height. At first approximation and ignoring for effects caused by the rotation of the Earth, winds blowing over a channel of constant depth bring about a slope of the water surface such that the resulting pressure gradient eventually balances the drag of the wind on the water surface

$$\frac{\partial \zeta}{\partial x} = \frac{\rho_a\, c_d u^2}{\rho\, Dg}, \tag{2.16}$$

where ζ represents sea surface height; x is the horizontal distance over which the wind is acting on the sea surface (fetch); ρ_a and ρ are the densities of the atmosphere and seawater, respectively; D denotes water depth; u is wind speed; and c_d and g are the drag coefficient and the acceleration of gravity, respectively. Equation (2.16) shows that the response of sea surface height to wind forcing basically depends on wind speed, fetch, and water depth. Figure 2.25 (see color section) illustrates these dependences. In general, surges get larger with increasing wind speeds and fetches. For instance, a moderate storm with Beaufort-8 wind speeds blowing over 100 km of water with a depth of 20 m causes a storm surge of about 36 cm. In the same geometrical configuration a Beaufort-12 storm would raise the sea surface by about 2 m. When the same Beaufort-12 storm is blowing over about 150 km of sea surface, surge heights would increase to about 3 m. Moreover, the response of sea surface height to wind forcing is inversely related to water depth. Under otherwise identical conditions the response is larger in shallower waters. When Beaufort-12 wind speeds are acting over about 100 km of fetch, storm surge heights would be about 2 m and 4 m for water depths of 20 m and 10 m, respectively. Figure 2.26 (see color section) also shows that, as a consequence of higher wind speeds, tropical cyclones are often associated with higher storm surges. For example, a Category-3 hurricane moving off the Texas Gulf Coast, where water depths are typically about 20 m, may lead to storm surge levels of about 3.25 m when a fetch of 50 km is adopted.

In the real world there are a number of processes that modify the simple picture provided by (2.16). When the rotation of the Earth is taken into account, the net transport of water caused by surface winds is not parallel to the wind direction, but to the right in the Northern Hemisphere and to the left in the Southern Hemisphere. This process is known as *Ekman transport* and it is particularly important when the wind blows parallel to the coastline. In shallow waters bottom friction opposes wind drag on the sea surface and reduces surge heights. Also interactions between tides and surges, caused by the presence of non-linear terms in the governing equations, may become important (e.g., Horsburgh and Wilson, 2007). Other effects comprise shoaling caused by changes in water depth, channeling effects in large estuaries or bays, presence of sea ice that may reduce effective fetch, or interactions with wind-generated waves at the sea surface. In the following some examples are provided.

Regional examples of storm surges. A global map of areas exposed to significant storm surge risk is shown in Figure 2.26. Mainly low-lying coastal areas influenced by tropical or extra-tropical cyclones are affected. For example, the North Sea, located on the European continental shelf between the British Isles and continental Europe, represents a storm-prone area which, especially in winter, is exposed to frequently occurring extra-tropical cyclones. Much of the land surrounding the North Sea is below or only slightly above mean sea level, making coastal flooding a particular serious hazard. The latter particularly applies to The Netherlands, Denmark, and Germany. For instance, in Germany the west coast of Schleswig-Holstein comprises about 553 km of coastline, 408 km of which are protected by dikes or other sea defenses (Ministerium für ländliche Räume, Landesplanung, Landwirtschaft und Tourismus des Landes Schleswig-Holstein, 2001). In the U.K. as well as in other countries surrounding the North Sea erosion caused by storm surges and storm waves represents a particular threat. The North Sea is located on the exit of the North Atlantic storm track and storms frequently travel eastward across its wide entrance in the north where they may set water into motion that propagates anti-clockwise along the North Sea coastline initially with very little resistance from bottom friction. This phenomenon is frequently referred to as an *external surge* in order to distinguish it from surges that are generated locally in the North Sea by the effects of wind and sea surface pressure. Figure 2.27 illustrates the effect. When a tide–surge model (see Section 3.5) for the North Sea is driven solely by astronomical forces, naturally quite large differences relative to the observed situation are obtained as the meteorological effects are ignored completely. In the example shown the differences between the real situation and a tide-only simulation

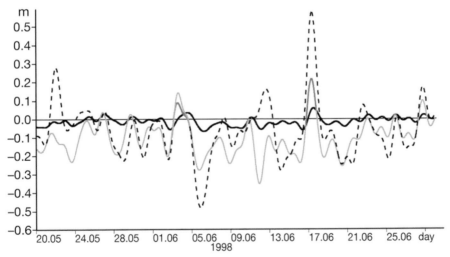

Figure 2.27. Differences between observed and simulated water levels at Cuxhaven in the southern North Sea for a tide-only simulation (dashed), for a simulation with tidal and local wind forcing (solid, light gray), and for a simulation driven by tides, local wind fields, and external surges (solid, black). External surges have been accounted for by the assimilation of observed water levels from a tide gauge near the model boundary (Aberdeen). After Weisse and Pluess (2006).

are larger than 50 cm around June 18. These differences are somewhat reduced when the effects of local wind and pressure fields are accounted for. When additionally external surges are taken into account in the model simulation, the differences between observations and model results are smallest.

Generally, external surges add to the effect of locally generated surges and may, under otherwise rather similar weather conditions, significantly raise observed water levels. An example is the North Sea storm surge that occurred on February 16–17, 1962 and caused severe damage in northern Germany and, particularly, in Hamburg. The meteorological situation observed over the North Sea on February 16–17 was rather similar to that about four days earlier with relatively moderate wind conditions of about Beaufort 9–10 (Koopmann, 1962). However, in the German Bight observed surge heights varied considerably between both events and observed high water levels differed by more than one meter (e.g., Weisse and Pluess, 2006). This difference is attributed to the influence of external surges that contributed to the second event, but were lacking for the first event (Koopmann, 1962). For much of the North Sea highest water levels were observed in January 1976. Here interaction with topography and the timing of the storm relative to the tide turned out to be important. Winds from the west hampered the outflow of water from the River Elbe during low tide and, after the passage of a cold front, winds turned towards more northwesterly directions which additionally pushed water into the estuary at the rising tide.

For shallow-water areas *tide–surge interactions* may become important and may modify the statistics of storm surges and high-water events. Tide–surge interaction refers to a modification in the propagation of surges by tides and *vice versa*. The phenomenon is attributed to non-linear terms in the governing equations, mainly to the shallow-water and bottom friction terms. A simple example of tide–surge interaction is the reduction of water depth at low tide which increases bottom friction and reduces propagation speed and the height of surges compared with a similar configuration at high tide. In reality, interaction may be more complex. For the River Thames tide–surge interaction has been studied and documented by Prandle and Wolf (1978). They showed that peak surges tend to occur before high and during rising tide while large storm surge peaks seldom, if ever, occurred at high tide. Similar result were obtained by Woodworth and Blackman (2002) who analyzed Liverpool tide–surge data since 1768 and by Horsburgh and Wilson (2007) for a number of different U.K. tide gauges.

The Baltic Sea with its surface of about $377,000 \, \text{km}^2$ represents one of the largest brackish inland seas. It has an average length of about 1,600 km and an average width of about 193 km. The maximum depth is about 460 m and the average depth is about 55 m. The Baltic Sea is characterized by an excess of precipitation and runoff over evaporation and is connected to the North Sea only via three narrow straits. As a consequence tidal elevations are negligible almost throughout the entire Baltic Sea and fetches to built-up large surges remain limited owing to the spatial configuration of the sea relative to the major paths of storms in the area. For instance, northeasterly winds of Beaufort 8–10 that are able to act on long fetches may cause storm surge heights of about 1.5 m to1.8 m at the southwestern shorelines as a response of the sea surface solely to wind forcing (Umweltministerium Mecklenburg-Vorpommern,

1997). There are, however, a number of other processes that may significantly contribute to extreme sea surface heights in the Baltic Sea in response to meteorological phenomena: long-lasting (several days) series of storms with wind directions preferably from the west to the southwest may push water through the straits that connect the Baltic Sea with the North Sea. As a consequence water levels in the entire Baltic Sea may be increased by 0.3 m to 0.5 m. Similarly, these storms push the water masses towards the eastern and northeastern coasts and raise the water levels there. Simultaneously, water levels at the southern and southwestern coasts drop. When the storms cease and wind speed drops, water flushes back towards the southern and southwestern coasts. The latter may raise water levels along the German and Danish coastline by up to 1 m (Umweltministerium Mecklenburg-Vorpommern, 1997).

Two exceptionally strong storm surge events in the Baltic Sea occurred in November 1872 and in January 2005. In November 1872, over the night of November 12 to 13 surge levels reached values of more than 3.50 m above mean sea level in the southwestern Baltic Sea and remained above 2 m for about 18 hours. Figure 2.28 shows annual surge maxima at Travemünde, Germany, where 3.30 m were observed. Obviously, the 1872 event was way beyond anything that has been recorded before and after. Another exceptionally strong surge hit the Estonian coast in January 2005. It was caused by Wind Storm Gudrun, which affected many areas in northern Europe between January 7 and 9, 2005. In the Estonian city of Pärnu, the observed storm surge associated with Gudrun reached 2.75 m above mean sea level and was the highest ever recorded. The exceptionally high surges were considered to be a result

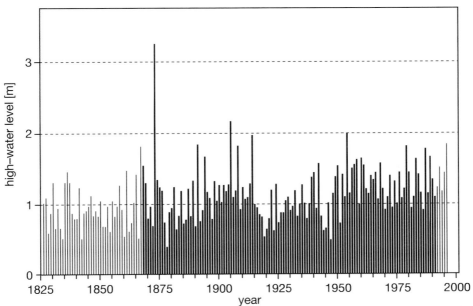

Figure 2.28. Observed annual maxima high-water levels in Travemünde, Schleswig-Holstein, Germany, since 1825.

of high wind speeds in combination with resonance effects of the wind storm moving over shallow water (Suursaar *et al.*, 2006).

The U.S. Gulf Coast and the Atlantic coast of Florida are other areas of enhanced storm surge risk (Figure 2.26). The Gulf of Mexico is almost completely landlocked and characterized by relatively warm water temperatures. Because of the latter, hurricanes entering the Gulf usually intensify on their path before making landfall. Because of the relatively shallow continental shelf, storm surges associated with hurricanes may be extreme. For instance, for a typical hurricane with wind speeds of about $40\,\mathrm{m\,s}^{-1}$ off the Texas Gulf Coast, where average depths and fetches are about $20\,\mathrm{m}$ and $100\,\mathrm{km}$, respectively, according to (2.16), corresponding surges will be in the order of $1\,\mathrm{m}$ to $2.5\,\mathrm{m}$ depending on how c_d is parameterized. Surges are generally larger in the forward right quadrant of a tropical storm (Figure 2.22) and their spatial scale is much smaller than that of extra-tropical surges. There may be, however, processes that complicate that simple figure. An example is provided by Hurricane Dennis, which hit the U.S. Gulf Coast on July 10, 2005 just east of Pensacola. Surges observed in Apalachee Bay, about $275\,\mathrm{km}$ east of Pensacola, were about $2\,\mathrm{m}$ to $3\,\mathrm{m}$; that is, about one meter higher than predicted. They could not be explained by the rather moderate winds in the area. An explanation for the unexpectedly high surges has been provided by Morey *et al.* (2006). Using a series of storm surge simulations under idealized and real conditions they showed that the additional 1-meter surge height could be explained by a topographic Rossby wave that traveled along the Florida Gulf Coast. While the storm was moving almost parallel to the coast it generated alongshore winds that produced onshore Ekman transport and a positive sea level anomaly at the coast. The latter finally traveled northward as a coastally trapped wave. From theoretical arguments Morey *et al.* (2006) estimated the speed of this wave to be comparable with that of the moving hurricane. As a result the wave continued to amplify on its path, leading to enhanced storm surge levels in Apalachee Bay.

There are many more areas in the world that are sensitive to storm surge risk (Figure 2.26). Examples comprise China, parts of northern Australia, and the Indian coastline. In all cases surge levels depend crucially on a number of local factors that may be more or less important, depending on details of the area and prevailing meteorological conditions.

2.5.2 Tides

Tides are regular periodic motions of the sea surface caused by gravitational forces. The only two astronomical bodies that exert sufficiently strong gravitational forces on the Earth to cause noticeable variations of the sea surface height are the Sun and the Moon. Because of its shorter distance to the Earth, the gravitational attraction of the Moon is about twice that of the Sun. When only the gravitational forces of the Moon are considered to act on an Earth covered entirely with water (an aqua planet) where the water is sufficiently deep and where the response to gravitational forces is instantaneous, the most simple concept of a tidal response, generally known as the *equilibrium tide*, is obtained. Here the response consists of the following components:

Sec. 2.5]   **Tides, storm surges, and mean sea level**   67

1. *A semidiurnal component that is controlled by the Earth's rotation.* The revolution of the Moon and the Earth around their joint center of mass causes two water bulges that are in line with an axis from the Moon to the Earth. Because of the Earth's rotation these two bulges travel over the Earth surface causing two high waters and two low waters a day at any given point under the influence of these two bulges. The period of this tidal component is only roughly semidiurnal as the rotation of the Moon around the Earth slightly increases the period to 12 hours 25 minutes. There is a small asymmetry in the height of the two bulges with the one more remote from the Moon being slightly smaller.

2. *A diurnal component that is controlled by lunar declination in combination with Earth's rotation.* Caused by lunar declination the maximum heights of the two semidiurnal bulges are not located on the Earth's equatorial plane but one in the Northern Hemisphere and the other in the Southern Hemisphere. As the Earth rotates, each point on the Earth's surface under the influence of the two semi-diurnal bulges will experience daily variations in tidal range.[22] Tidal ranges are increased when the point is under the influence of the bulge that has its maxima in the same hemisphere and are reduced when Earth's rotation has brought the point under the influence of the other bulge roughly half a day later. As lunar declination varies with a period of 27.2 days (nodal month), asymmetry varies with the same period. When lunar declination is large, asymmetry is also large and may, in extreme cases, almost suppress the tidal range under one bulge while largely increasing it under the other.

3. *Long-term components.* These components have periods of a month and longer and are associated with changes in the Earth–Moon distance and declination (e.g., Pugh, 1987). One of the longest periods (of about 18.6 years) is caused by variation of the inclination of the plane in which the Moon orbits the Earth (nodal cycle).

A similar equilibrium tide of smaller amplitude is found when the gravitational forces of the Sun are analyzed in a similar way. When the forces of both the Sun and the Moon are considered, additional modifications occur. The most prominent phenom-enon is the so-called spring–neap cycle in which the tidal range of the semidiurnal component controlled by the Moon is modified by interaction with gravitational forces from the Sun. Here the range of the semidiurnal tide is varied such that it increases and decreases over a period of 14 days. Maximum ranges typically occur when the Sun and the Moon are in line (spring tides), minimum ranges occur when both bodies are in quadrature (neap tides). Semidiurnal tides controlled by the Moon with a superimposed 14-day spring–neap cycle caused by the additional forces exerted by the Sun represent the dominant tidal pattern for the greater part of the world oceans.

The response described so far has been that of equilibrium tides. In the real world the situation is more complex. The existence of continents and other barriers affects the propagation of tides. As tides can be considered very long waves with wavelengths

---

[22] That is, the difference between subsequent high and low waters.

of up to half the circumference of the Earth in the case of a lunar semidiurnal tide these waves represent shallow water waves and they experience the influence of the bottom even within the deep ocean. The propagation speed c of shallow water waves is given by $c = \sqrt{gh}$ where h represents water depth and g the gravitational acceleration. Given these circumstances the propagation speed of a tidal wave is too small to follow the Moon even in the deep ocean. For example, the speed of a shallow water wave at about 3,000 m depth would be approximately $170\,\mathrm{m\,s^{-1}}$. At the equator, a speed of roughly $460\,\mathrm{m\,s^{-1}}$ would be required to follow the Moon's orbit. As a result, tides in the real world are more complex than those obtained from the simple consideration of gravitational forces. An example of tides in the real world is given in Figure 2.29 (see color section) for the lunar semidiurnal component. Tidal waves tend to rotate around so-called *amphidromic points* for which the tidal range is almost zero. Amphidromic points occur as a result of the combined effects of gravitational forces, the Coriolis force, and interaction of the tidal wave with the ocean basins, bathymetry, and the coast. The direction of rotation is mainly clockwise in the Southern Hemisphere and anticlockwise in the Northern Hemisphere. Tidal ranges are usually largest near the coasts.

A noticeable direct response to gravitational forces is only found in the open oceans. Shelf seas are usually too small to experience a significant direct impact. Nevertheless, tidal ranges here are often large especially in semi-enclosed basins. The latter is a result of tides propagating into the basins from the open ocean sometimes in combination with tidal resonance. Tidal resonance occurs when the time it takes the tidal wave to propagate into the basin, to reflect, and to travel back is approximately the same as that between two subsequent high waters. One of the most prominent examples of tidal resonance is found in the Bay of Fundy, Canada, where tidal ranges in the order of 13 m are observed (e.g., Pugh, 2004). Figure 2.30 (see color section) shows another example of tidal patterns in a continental shelf sea—here the North Sea. The progress of tidal waves can be inferred from cotidal lines. A tidal wave enters the North Sea from the Atlantic and travels southward along the U.K. coastline with tidal ranges mostly between 3 m and 4 m near the coast. The wave then propagates along the Dutch, the German, and the Danish coasts with slightly smaller amplitudes. Another tidal wave enters the North Sea from the southwest through the English Channel. There are three amphidromic points, one in the southwest, one in the southern North Sea, and one off the Norwegian coast. Tides in the North Sea have noticeable amplitudes mainly because of its open northern boundary through which tidal waves from the Atlantic can propagate. Shelf seas and basins with comparable spatial extent but with less distinct connections to the open oceans have considerably smaller tidal ranges that are sometimes hardly noticeable. Examples are the Baltic and the Mediterranean Seas.

Mathematically, the response of the ocean to gravitational forces can be described by superposition of a finite number of harmonic constituents, the periods of which are determined from astronomical arguments. The constituents are then fitted to observed data to determine local amplitudes and phases that may subsequently be used for tidal prediction. The procedure is generally known as *harmonic analysis* and is described, for instance, in Pugh (2004).

Tides may interact with storm surges (see Section 2.5.1) but tidal patterns may also change in response to changes in mean sea level or local construction works. An example is given in Kauker (1999) who studied the response of tidal patterns in the North Sea to changes in mean sea level using a numerical ocean model. For an increase in mean sea level of one meter the principal lunar (M2) cotidal lines show small shifts of amphidromic points and tidal elevation is increased by about 1 cm to 4 cm for most of the southeastern North Sea. Similar experiments were later carried out by Pluess (2006) for a number of different sea level scenarios and, at the same time, using a model with greatly improved spatial resolution near the coasts and in the estuaries. He found that near coasts and within estuaries the effect may be significantly enhanced.

As it is the combined effect of tides, surges, and mean sea level that is responsible for high water levels and coastal flood risk, changes in all three components and their interaction must be studied when long-term changes are considered. So far, mainly the effects of mean sea level changes have been emphasized, mostly because low-lying coastal areas appear to be vulnerable to even small changes in mean sea level and because global mean sea level rise is regarded as one of the more certain outcomes of global warming (Houghton *et al.*, 2001).

2.5.3 Mean sea level

Mean sea level, as introduced in (2.13), is defined as the average height of the sea surface relative to a fixed benchmark. Averaging should be performed over a sufficiently long period to account for and to remove the effects of tides and surges on sea surface height. In the United States, for example, averaging is performed over a 19-year period to account for even the longest tidal cycles. In the undisturbed case, mean sea level would adjust to an equipotential surface[23] known as the geoid. In particular, this would be the case if there were

— no tidal forces by the Moon or the Sun,
— no spatial density differences in the oceans caused by temperature or salinity variations, the latter also includes spatial variations in evaporation and precipitation,
— no atmospheric pressure variations along the geoid surface, and
— no winds or currents.

In reality, mean sea level therefore deviates from the geoid. While deviations of the geoid from the ellipsoid may be up to about 100 m, deviations in mean sea level from the geoid are two orders of magnitude smaller and may reach up to about one meter.

---

[23] A surface along which a particle can move without any work against the force of gravity. On a rotating Earth with homogeneous density distribution and no other forces the shape of this surface would be an ellipsoid. In reality, there are density variations between the solid Earth and the oceans that lead to deviations from the ellipsoidal shape. The deviations are in the order of 10 m to 30 m at the poles and may reach up to 100 m south of India (Pugh, 2004).

The lowest mean sea level is found around Antarctica while highest values are observed in the tropics. This is a result of ocean density variations caused by temperature differences, with higher temperatures in the tropics and colder temperatures around Antarctica. As the tropical Pacific is somewhat warmer than the tropical Atlantic, there is a difference in mean sea level of about 20 cm between both ends of the Panama Canal (Pugh, 2004).

Changes in mean sea level may arise from several factors. We principally distinguish between geological factors that may have impact on both the land and the oceans, and non-geological factors that may change the volume of oceans and hence may affect mean sea level. On land, the geological factors that may contribute to mean sea level changes are *isostatic adjustment* and *tectonic movements*. Isostatic adjustment refers to widespread vertical land movements that are the result of varying surface loads caused by the growth/decay and advance/retreat of ice sheets in the course of glacial–interglacial cycles. The vertical land movements caused by isostatic adjustment show a spatially and temporally highly complex pattern. For instance, for the Baltic Sea area isostatic adjustment varies between about -1 mm yr^{-1} and $+9$ mm yr^{-1} relative to the coastline (Ekman, 1996). Sea level changes caused by isostatic adjustment are referred to as *isostatic sea level changes*. Tectonic movements may occur as a result of rather rapid changes such as earthquakes or volcanic eruptions, but also from plate tectonic movements. Similarly, they result in vertical land movement that affects mean sea level. Since major tectonic activity is observed mainly at plate boundaries, which frequently also represent continental or island boundaries, sea level signals at many tide gauges do have contributions from tectonic movements. The geological factors described may not only change the land surface and elevation, but also the shape of the ocean basins and as a consequence mean sea level.

Apart from geological changes there are also non-geological factors that may change mean sea level. These factors generally affect the *volume* of water in the oceans, and the sea level changes caused by these factors are referred to as *eustatic changes*. Changes in the volume of ocean water may occur from changes of the mass or the density of seawater. The mass of water in the ocean may change as a result of mass exchange with glaciers, ice caps, or terrestrial storage such as permafrost, ground water, or surface water. Changes in ocean water density[24] may occur primarily from water temperature changes and to a smaller extent from salinity changes. Generally, when globally averaged ocean temperature increases, density decreases, and sea level rises as a result of the increase in ocean volume. The latter is known as *thermal expansion* or *steric sea level rise* and is expected to provide the dominant contribution to sea level rise in the course of anthropogenic climate change, at least within the next few decades. Later, melting of large continental ice sheets such as Greenland and Antarctica may become more important. Expected and observed sea level changes are treated in detail in Section 5.5.

Eustatic sea level changes are not uniform around the globe. Their geographical distribution depends on alterations in the ocean density structure with corresponding

---

[24] At constant mass.

effects on ocean circulation. Long-term changes in the spatial distribution of atmospheric surface pressure patterns may also lead to regional sea level changes caused by the inverse barometric effect (see Section 2.5.1). Similarly, temperature variability in the atmosphere and the oceans may lead to decadal and inter-annual variations of mean sea level. An example is provided in Church *et al.* (2005) who developed and tested a hypothesis that part of the recently observed sea level rise may be due to a recovery of sea level from the cooling effects of the Mt. Pinatubo volcanic eruption in 1991.

Studying sea level trends, particularly within the context of anthropogenic climate change, requires time series that are long compared with observed variability. In addition, nearly global coverage and homogeneous[25] time series are required to avoid sampling of re-distribution or other effects. There are principally two data sources from which changes in mean sea level can be derived: tide gauges and satellites. So far, both sources have advantages and limitations in providing data suitable to study long-term changes in mean sea level. Compared with satellites, tide gauge data are available for relatively long time periods. However, the spatial distribution of tide gauges is not optimal for analysis of long-term changes. There are more tide gauges along the coasts than in the open ocean, more in the Northern Hemisphere than in the Southern Hemisphere, and more tide gauges in highly developed countries. While the records obtained from some tide gauges are long compared with those from satellites, the majority of records are less than 30 years long (Pugh, 2004) which is still rather short for analysis of long-term sea level changes. There are a few records that are more than 100 years long, but most of them have been initially placed for port operations and are influenced by local changes such as those caused by local water works. Compared with tide gauges, satellite data show a better spatial coverage. At the moment the length of the record is, however, still rather limited but satellite measurements will increasingly become important in the future when data for longer time periods will become available. In addition, there are some difficulties—in particular, with the correction for instrument effects such as satellite bias and drift—and a high-quality tide gauge network is needed to correct for such effects. Interpretation of observed sea level changes therefore remains a difficult task (some aspects of which will be addressed in Section 5.5).

2.6 SUMMARY

Wind storms and their marine companions—wind waves and storm surges—may have significant impacts on coastal areas and operations at sea. In the mid-latitudes, extra-tropical cyclones, feeding on the temperature contrast between cold polar and temperate air masses of tropical origin at the polar front, represent the most prominent phenomenon. These cyclones are usually associated with high wind speeds and propagate along preferred trails, so-called storm tracks. The latter are generally more pronounced over the oceans with centers over the mid-latitude Pacific, Atlantic, and

---

[25] See Section 4.2.1.

72 **Marine weather phenomena** [Ch. 2

Southern Ocean. In the tropics, tropical cyclones are the most relevant phenomenon responsible for high impacts. The impacts are caused by high wind speeds, severe storm surges, extreme ocean wave heights, or heavy precipitation associated with these storm systems, or a combination of these factors. Contrary to mid-latitude storms, tropical cyclones are non-frontal low-pressure systems that are maintained mainly by the extraction of latent heat from the oceans at high temperatures and by heat export to the upper troposphere at low temperatures. Consequently, tropical cyclones represent a warm season phenomenon, which, in each hemisphere, occurs only during the summer months. Usually, activity peaks in late summer when upper-ocean temperatures are warmest.

Apart from tropical and extra-tropical cyclones there exist a number of other phenomena that may cause high wind speeds at sea. At high latitudes the most prominent features are polar lows. Polar lows are small but intense cyclones that form preferably when cold polar air is advected over warmer water (Rasmusson and Turner, 2003). The latter often happens at sea ice margins. Horizontal scales of polar lows range from several tens to several hundreds of kilometers and because of their strong winds and heavy precipitation these storms are sometimes referred to as "Arctic hurricanes" (e.g., Glickman, 2000). Between the Equator and about 50° latitude in each hemisphere, storms referred to as subtropical cyclones may occur. These storms are found in regions of weak to moderate horizontal temperature contrast and extract their energy from baroclinic instabilities as well as from convective redistribution maintained by warm sea surface temperatures (Glickman, 2000). As such these cyclones do have characteristics of both tropical and extra-tropical cyclones and may sometimes convert to tropical or extra-tropical cyclones, respectively. Similarly, they represent an intermediate stage for tropical cyclones when undergoing extra-tropical transition (e.g., McTaggart-Cowan et al., 2007). There are a number of other meteorological phenomena associated with high wind speeds. These include, for instance, tornados or thunderstorms. The latter generally have much smaller scales and usually occur in connection with the phenomena discussed above.

Over the oceans and in coastal areas high wind speeds may cause variations in sea surface height that may lead to serious impacts. For the open ocean wind-generated waves are most relevant. These are short-scale ocean surface waves caused by the drag of the sea surface. The height of such waves depends crucially on the duration of high wind speeds, wind speed itself, and fetch (i.e., the amount of open water available for the wind to build up waves). Under severe tropical and extra-tropical cyclones, significant wave heights of more than 16 m have been reported (e.g., Cardone et al., 1996). While significant wave height refers to the average height of the highest 33% of the waves within a given time interval (usually 20 minutes) individual waves may be much higher. Under particular circumstances they may become especially high and steep (freak or rogue waves), which represents a particularly severe hazard for shipping and offshore operations.

For coastal areas, storm surges constitute a particular threat. Storms surges are a companion of low-pressure systems and result from high wind speeds pushing water towards the coast, and from an increase in sea surface height caused by low atmospheric pressure within the cyclone (inverse barometric effect). While local characteristics—such as water depth, shape of the ocean basin, or configuration of the surrounding land masses—play a crucial role in the specific characteristics of storm surges at specific places, the most devastating storm surges are usually associated with tropical cyclones. Here spatially confined but intense storm surges occur that have typical time scales of a few hours hours and may affect coastal stretches of up to about 200 km (von Storch and Woth, 2008). Extra-tropical surges, on the other hand, have much larger spatial dimensions of up to several hundreds of kilometers, and longer time scales on the order of one day. Storm surges interact with and add to the sea level variations caused by tides and mean sea level changes.[26] As such, long-term variations in any of these factors may alter the risk of coastal flooding. So far, mainly variations in mean sea level have been considered when changing coastal flood risks have been addressed.

2.7 REFERENCES

Baxter, P. (2005). The East Coast Great Flood, 31 January–1 February 1953: A summary of the human disaster. *Phil. Trans. R. Soc.*, **363**, 1293–1312.

Blackmon, M. (1976). A climatological spectral study of the 500 mb geopotential height of the Northern Hemisphere. *J. Atmos. Sci.*, **33**, 1607–1623.

Camargo, S.; A. Sobel; A. Barnston; and K. Emanuel (2007). Tropical cyclone genesis potential index in climate models. *Tellus*, **59A**, 428–443, doi: 10.1111/j.1600–0870.2007.00238.x.

Cardone, V.; R. Jensen; T. Resio; V. Swail; and A. Cox (1996). Evaluation of contemporary ocean wave models in rare extreme events: The "Halloween Storm" of October 1991 and the "Storm of the Century" of March 1993. *J. Atmos. Oceanic Technol.*, **13**, 198–230.

Christoph, M.; U. Ulbrich; and U. Haak (1995). Faster determination of the intraseasonal variability of storm tracks using Murakami's recursive filter. *Mon. Wea. Rev.*, **123**, 578–581.

Church, J.; N. White; and J. Arblaster (2005). Significant decadal-scale impact of volcanic eruptions on sea level and ocean heat content. *Nature*, **483**, 74–77, doi: 10.1038/nature04237.

Didenkulova, I.; A. Slunyaev; E. Pelinovsky; and C. Kharif (2006). Freak waves in 2005. *Nat. Hazards and Earth Syst. Sci.*, **6**, 1107–1115, doi: www.nat-hazards-earth-syst-sci.net/6/1007/2006.

Ekman, M. (1996). A consistent map of postglacial uplift of Fennoscandia. *Terra Nova*, **9**, 158–165.

Emanuel, K. (1986). An air–sea interaction theory for tropical cyclones, Part I: Steady-state maintenance. *J. Atmos. Sci.*, **43**, 585–604.

Emanuel, K.; and D. Nolan (2004). Tropical cyclone activity and global climate. In: *Proc. 26th Conference on Hurricanes and Tropical Meteorology*. American Meteorological Society, Miami, FL.

Gerritsen, H. (2005). What happened in 1953? The big flood in the Netherlands in retrospect. *Phil. Trans. R. Soc.*, **363**, 1271–1291.

---

[26] Changes on very long time scales associated, for instance, with vertical land movements or with thermal expansion of the ocean under global warming.

74 **Marine weather phenomena** [Ch. 2

Glickman, T. (Ed.) (2000). *Glossary of Meteorology*. American Meteorological Society, Boston, MA, Second Edition.

Gönnert, G.; S. Dube; T. Murty; and W. Siefert (2001). Global storm surges. *Die Küste*, 63. ISBN 978-3-804-21054-6, 624 pp.

Gray, W. (1968). A global view of the origin of tropical disturbances and storms. *Mon. Wea. Rev.*, **96**, 669–670.

Greenslade, D. (2001). A wave modelling study of the 1998 Sydney to Hobart yacht race. *Aust. Met. Mag.*, **50**, 53–63.

Hartmann, D. (1994). *Global Physical Climatology*. Academic Press, London, 411 pp.

Haver, S. (2000). *Evidence of the Existence of Freak Waves*, Technical Report. Statoil, E&P Norway, Stavanger, Norway.

Haver, S.; and O. Andersen (2000a). *Freak Waves: Myth or Reality?* Technical Report. Statoil, E&P Norway, Stavanger, Norway.

Haver, S.; and O. Andersen (2000b). Freak waves rare realizations of a typical population or typical realizations of a rare population? In: *ISOPE-2000, Seattle.*

Horsburgh, K.; and C. Wilson (2007). Tide–surge interaction and its role in the distribution of surge residuals in the North Sea. *J. Geophys. Res.*, **112**, C08003, doi: 10.1029/2006JC004033.

Horstmann, J.; D. Thompson; F. Monaldo; F. Iris; and H. Graber (2005). Can synthetic aperature radars be used to estimate hurricane force winds? *Geophys. Res. Lett.*, **32**, L22801, doi: 10.1029/2005GL023992.

Houghton, J.; Y. Ding; D. Griggs; M. Noguer; P. van der Linden; X. Dai; K. Maskell; and C. Johnsn (Eds.) (2001). *Climate Change 2001: The Scientific Basis. Contribution of Working Group I to the Third Assessment Report of the Intergovernmental Panel on Climate Change.* Cambridge University Press, Cambridge, U.K. ISBN 0521 01495 6, 881 pp.

Kallberg, P.; P. Berrisford; B. Hoskins; A. Simmons; S. Uppala; S. Lamy-Thépaut; and R. Hine (2005). *ERA-40 Atlas*, Technical Report 19. ECMWF, Reading, U.K.

Kauker, F. (1999). Regionalization of climate model results for the North Sea. Ph.D. thesis, University of Hamburg, Hamburg, Germany. Available as GKKS Rep. 99/E/6 from GKSS-Forschungszentrum, Geesthacht, Germany.

Kjeldsen, S. (1997). Examples of heavy weather damages caused by giant waves. *Techno Marine, Bull. Soc. Naval Architects Japan*, **10**, 24–28.

Koopmann, G. (1962). Die Sturmflut vom 16./17. Februar 1962 aus ozeanographischer Sicht. *Die Küste*, **10**, 55–68 [in German].

Kraus, H.; and U. Ebel (2003). *Risiko Wetter*. Springer-Verlag, Berlin [in German]. ISBN 3-540-00184-0, 250 pp.

Landsea, C. (2008). Why doesn't the South Atlantic Ocean experience tropical cyclones? Available at *http://www.aoml.noaa.gov/hrd/tcfaq/G6.html*, last accessed March 25, 2008.

Liu, K. (2004). Paleotempestology: Principles, methods, and examples from Gulf Coast lake sediments. In: R. Murnane and K. Liu (Eds.), *Hurricanes and Typhoons: Past, Present and Future*. Columbia University Press, pp. 13–57.

Longuet-Higgins, M. (1952). On the statistical distribution of the wave heights of sea waves. *J. Marine Res.*, **11**, 245–266.

Maue, R. (2009). Northern Hemisphere tropical cyclone activity. *Geophys. Res. Lett.*, **36**, L05805, doi: 10.1029/208GL035946.

McTaggart-Cowan, R.; L. Bosart; J. Gyakum; and E. Atallah (2007). Hurricane Katrina (2005), Part I: Complex life cycle of an intense tropical cyclone. *Mon. Wea. Rev.*, **135**, 3905–3926, doi: 10.1175/2007MWR1875.1.

Ministerium für ländliche Räume, Landesplanung, Landwirtschaft und Tourismus des Landes Schleswig-Holstein (2001). Generalplan Küstenschutz-Integriertes Küstenschutzmanagement in Schleswig-Holstein. Available at *http://www.schleswig-holstein.de/Unwelt Landwirtschaft/DE/WasserMeer/09–KuestenschutzHaefen/02–GeneralplanKuestenschutz/ ein–node*, last accessed May 19, 2009 [in German].

Monbaliu, J.; and A. Toffoli (2003). Regional distribution of extreme waves. In: *Proceedings of MAXWAVE Final Meeting, October 8–10, Geneva, Switzerland*, 10 pp. Available at *http:// coast.gkss.de/projects/maxwave/meetings/Genf/minutes_genf.htm*

Morey, S.; S. Baig; M. Bourassa; D. Dukhovskoy; and J. O'Brien (2006). Remote forcing contribution to storm-induced sea level rise during Hurricane Denis. *Geophys. Res. Lett.*, **33**, L19603, doi: 10.1029/2006GL027021.

Müller-Navarra, S.; I. Bork; J. Jensen; C. Koziar; C. Mudersbach; A. Müller; and E. Rudolph (2006). Modellstudien zum Sturmflut und zum Hamburg Orkan 1962. *Hansa*, **43**, 72–88 [in German].

Pezza, A.; and I. Simmonds (2005). The first South Atlantic hurricane: Unprecedent blocking, low shear and climate change. *Geophys. Res. Lett.*, **32**, L15712, doi: 10.1029/ 2005GL023390.

Pichler, H. (1997). *Dynamik der Atmosphäre*. Spektrum Akademie-Verlag, Heidelberg [in German]. ISBN 3-8274-0134-8.

Pluess, A. (2004). Nichtlineare Wechselwirkung der Tide auf Änderungen des Meeresspiegels im Übergangsbereich Küste/Ästuar am Beispiel der Elbe. In: G. Gönnert, H. Grassl, D. Kellat, H. Kunz, B. Probst, H. von Storch, and J. Sündermann (Eds.), *Proceedings of Workshop Klimaänderung und Küstenschutz, November 29–30, Hamburg, Germany*. Available at *http://coast.gkss.de/staff/storch/pdf/kliky.proc.0411.pdf* [in German].

Prandle, D.; and J. Wolf (1978). The interaction of surge and tide in the North Sea and the River Thames. *Geophys. J. Roy. Astron. Soc.*, **55**, 203–216.

Pugh, D. (1987). *Tides, Surges and Mean Sea Level*. John Wiley & Sons, New York, 472 pp. ISBN 047191505x.

Pugh, D. (2004). *Changing Sea Levels: Effects of Tides, Weather and Climate*. Cambridge University Press. ISBN 9780521532181.

Rasmusson, E.; and J. Turner (2003). *Polar Lows: Mesoscale Weather Systems in the Polar Regions*. Cambridge University Press.

Rosenthal, W.; S. Lehner; H. Dankert; H. Güenther; K. Hessner; J. Horstmann; A. Niedermeier; J. Nieto-Borge; J. Schulz-Stellenfleth; and K. Reichert (2003). Detection of extreme single waves and wave statistics. In: *Proceedings of MAXWAVE Final Meeting, October 8–10, Geneva, Switzerland*, 6 pp. Available at *http://coast.gkss.de/projects/ maxwave/meetings/Genf/minutes_genf.htm*

Sand, S.; N. Ottesen-Hansen; P. Klinting; O. Gudmestad; and M. Sterndorff (1990). Freak wave kinematics. In: A. Tørum and O. Gudmestad (Eds.), *Water Wave Kinematics*. Kluwer Academic, Dordrecht, The Netherlands.

Shepherd, J.; and T. Knutson (2007). The current debate on the linkage between global warming and hurricanes. *Geography Compass*, **1**, 1–24, doi: 10.1111/j.1749–8198.2006.00002.x.

Simpson, R.; and H. Riehl (1981). *The Hurricane and Its Impact*. Louisiana State University Press, 398 pp. ISBN 978-0807106884.

Sönnichsen, U.; and J. Moseberg (2001). *Wenn die Deiche brechen: Sturmfluten und Küstenschutz an der schleswig-holsteinischen Westküste und Hamburg*. Husum Druck-und Verlagsgesellschaft, Husum, Germany [in German]. ISBN 3-88042-690-2.

76 Marine weather phenomena [Ch. 2

Soomere, T.; A. Behrens; L. Tuomi; and J. Nielsen (2008). Wave conditions in the Baltic Proper and in the Gulf of Finland during Windstorm Gudrun. *Nat. Hazards Earth Syst. Sci.*, **8**, 37–46.

Suursaar, U.; R. Kullas; M. Otsmann; I. Saaremäe; J. Kuik; and M. Merilain (2006). Cyclone Gudrun in January 2005 and its hydrodynamic consequences in Estonian coastal waters. *Boreal Environ Res.*, **11**, 143–159.

Sverdrup, H.; and W. Munk (1947). *Wind, Sea and Swell: Theory of Relations for Forecasting*, Publication No. 601. U.S. Navy Hydro Office, Washington, D.C., 44 pp.

Umweltministerium Mecklenburg-Vorpommern (1997). Küstenschutz in Mecklenburg-Vorpommern, Online Brochure. Available at *http://www.um.mv-regierung.de/ kuestenschutz/bschutz/* [in German].

van Dorn, W. (1994). *Oceanography and Seamanship*. Cornell Maritime Press, Second Edition.

van Heerden, I.; and M. Bryan (2006). *The Storm*. Viking, Penguin Group. ISBN 0670037818.

Vecchi, G.; and B. Soden (2007). Increased tropical Atlantic wind shear in model projections of global warming. *Geophys. Res. Lett.*, **34**, L08702, doi: 10.1029/2006GL028905.

von Storch, H.; and K. Woth (2008). Storm surges, perspectives and options. *Sustainability Science*, **31**, doi: 10.1007/s11625-008-0044-2.

Weisse, R.; and A. Pluess (2006). Storm-related sea level variations along the North Sea coast as simulated by a high-resolution model 1958–2002. *Ocean Dynamics*, **56**, 16–25, doi: 10.1007/s10236-005-0037-y, online 2005.

WMO (1998). *Guide to Wave Analysis and Forecasting*, WMO Vol. 702. World Meteorological Organization, Geneva, Switzerland. ISBN 92-63-12702-6.

Wolf, J.; and R. Flather (2005). Modelling waves and surges during the 1953 storm. *Phil. Trans. R. Soc.*, **363**, 1359–1375, doi: 10.1098/rsta.2005.2572.

Woodworth, P.; and D. Blackman (2002). Changes in extreme high waters at Liverpool since 1768. *Int. J. Climatol.*, **22**, 697–714.

Yasuada, T.; N. Mori; and S. Nakayama (1998). Characteristics of giant freak waves observed in the Sea of Japan. In: *Proc. Waves '97 Ocean Wave Measurement and Analysis, Virginia Beach, 1997*, Vol. 2.

3

Models for the marine environment

3.1 INTRODUCTION

Models are widely used in environmental sciences. However, the word *model* covers a much broader range than usually recognized by the user. In different areas of science, different meanings prevail and are considered to be correct. These differences can cause much confusion and problems for interdisciplinary cooperation. What a *model* or even a *good model* constitutes is a matter of social or cultural agreements within a wider or broader scientific field.[1] When referring to models a large variety of different concepts is generally meant, ranging from simple analogs like maps, to idealizations, conceptualizations, huge miniaturizations and, particularly in climate science, to a mathematically constructed substitute reality (von Storch, 2001).

Models are supposed to describe, explain, or at least reflect reality. But they are not identical to reality. Models are *smaller*, *simpler*, and *closed* in contrast to reality, which is always *open*. Here

1. *Smaller* means that only a limited number of the infinite number of real processes can be accounted for. In the case of an atmospheric model or an oceanic model, the unavoidable discretization means that from the overall range of scales only a limited interval can be accounted for. A global atmospheric model describes planetary waves and cyclones, but not boundary layer turbulence in any detail.
2. *Simpler* means that the description of the considered processes is simplified. For instance, when air is blowing across the surface of the oceans, surface friction is maintained by a cascade of small-scale turbulent eddies, which cannot be

---

[1] For a broader discussion of this field see publications dealing with the philosophy of science (e.g., Okasha, 2002) or, more specifically, of modeling (e.g., Müller and von Storch, 2004; Hesse, 1970) and references therein.

78 **Models for the marine environment** [Ch. 3

resolved by any numerical model. Instead their *overall* effect is described by *parameterizations* (see Section 1.2).

3. *Closed* means that models are integrated with a limited number of completely specified external forcing functions. For instance, in storm surge simulations wind force and tides are taken into account. The presence of oil slicks, quickly changing morphodynamics, or river runoff are usually neglected. This implies that the answer of a model may be "right" because the model is "correct" or because model errors are balanced by unaccounted external influences (Oreskes *et al.*, 1994).

The purpose of all models is to construct additional knowledge, more knowledge than what has been used to build the models. Using Hesse's language (Hesse, 1970), model and reality have properties, some of which are known to be shared by both and others that are known not to be shared. They are named *positive* and *negative analogs*. To determine these positive and negative analogs is to *validate* the model. But there are other properties, for which it is *unknown* whether they represent joint properties or not. These are the *neutral analogs*. Using a model means to assume that certain neutral analogs are actually positive ones. In, for example, the most simple case a storm surge model is validated with data from a number of tide gauges, but it is assumed implicitly that it will also provide reasonable results for places for which no measurements exist. Validation thus does not provide any new knowledge about reality but only about the model. Constructive use of the model (i.e., the construction of new knowledge about the real world) means using neutral analogs as positive ones (Müller and von Storch, 2004).

Models are always constructed for a *purpose*; namely, for exploiting certain neutral analogs. Thus, the expression "a model of Y" is not helpful. Instead the formulation "a model for describing property X of Y" is more appropriate. Speaking of "a model of the North Sea" makes little sense, but the terminology "a model describing variations of the circulation of the North Sea on time scales of days to weeks" is well defined. Unfortunately, the concept of "a model of Y" keeps many scientists busy; in these cases, the modeling problems are reduced to developing efficient algorithms and to validation. Certainly, both are important tasks, but they are nevertheless of secondary importance compared with the problem of designing a model for a specific purpose.

Two fundamentally different types of mathematical models are mainly used in meteorology, oceanography, and, more generally, in climate research:

1. *Quasi-realistic models* are supposed to constitute a reality substitute, within which otherwise impossible experiments can be conducted. A representative of this type are general circulation models (GCMs) of the atmosphere and the ocean (see Section 3.3.1). They are used, for instance, to probe the impact of enhanced greenhouse gas concentrations in the future, an experiment which cannot be carried out directly with the environment several hundreds times, as would be required in classical science to reach an inference.

2. *Cognitive models*, in contrast to quasi-realistic models, are highly simplified and idealized. Because of their reduced complexity such models constitute "knowledge" and are usually used to explain or conceptualize phenomena. Examples are zero-dimensional energy balance models (e.g., Crowley and North, 1991), but also the butterfly of Lorenz (1963), the stochastic climate model of Hasselmann (1976) (see Section 1.4.1), or the delayed action oscillator by Suarez and Schopf (1988).

In the following we discuss only quasi-realistic models. In Section 3.2 we first discuss some more general aspects. We then introduce a number of quasi-realistic models relevant for applications to the marine environment. These comprise models of the climate system on the globe and in limited areas (Section 3.3), models of ocean surface waves (sea states) (Section 3.4), and tide–surge models (Section 3.5). In our discussion, we limit ourselves to those aspects that are relevant for applications to the marine *climate*; that is, the detailed development of the geophysical properties extending over months to decades of years by generating realistic sequences of marine weather.

3.2 QUASI-REALISTIC MODELING

The main purpose of a quasi-realistic model is to provide scientists with an experimental and a simulation tool. As such, it is as complex as possible. Ideally, quasi-realistic models generate numbers as detailed and complex as the real world, within the limits of the spatial and temporal resolution of the model. Quasi-realistic models do not provide intuitive understanding, but provide large quantities of numbers, which are not immediately useful. Thus, constructive use of quasi-realistic models requires also a suitably prepared evaluation strategy for analysis of the generated data. Without such an analysis, modeling with quasi-realistic models hardly helps to better understand, examine, or predict marine environmental systems. Examples of quasi-realistic models are climate models, models of the hydrodynamics of marginal seas, or models of the sea state (wind wave models).

Quasi-realistic models describe the hydrodynamics and (often) the thermo-dynamics of the system in some detail and quantify many related processes in the form of parameterizations. But such models are considerably simpler than reality even if they strive for maximum complexity. In particular, their limitations comprise

1. *They describe only part of the reality*. For instance, a model describing the movement of a water body under the influence of winds and tides in a vertically homogeneous manner (vertically integrated model) is often sufficient to simulate storm surges in regional seas, whereas the effect of vertical stratification and baroclinic effects are disregarded. Another example refers to ocean surface waves, the statistics of which are often referred to as the sea state (see Section 2.4). Advanced wave models assume that the distribution of the surface of the

80 Models for the marine environment [Ch. 3

water may be understood as a statistically homogenous ensemble of propagating waves. Then, the model describes the temporal and spatial development of the spectrum of these wave components and not the development of the individual waves.

2. *The models cannot be verified.* We cannot with certainty conclude that the model is producing the "right" numbers because of the "right" dynamics. What we can do is to validate the models by finding that the simulated numbers are consistent with the observations. We may add to the credibility of the model by analyzing the dynamical system and assuring that all first-order, or even all second-order, processes are adequately accounted for. However, even the most comprehensive validation does not provide certainty about the skill of the models when they are applied in constructive mode; that is, when statements are made beyond the empirical evidence the models have been validated against. However, the purpose of quasi-realistic models is just that they are used beyond these limits by assuming that some neutral analogs represent positive ones (see Section 3.1).

In the following sections we will describe climate, wind wave, and tide–surge models in more detail. Before doing so, we have to emphasize that environmental issues related to the ocean, the atmosphere, and the climate exhibit a number of specifics, which make the field different from, for instance, classical physics (e.g., Navarra, 1995):

1. *The impossibility of conducting laboratory experiments.* Here, following *Encyclopedia Britannica*, we understand an experiment as "an operation carried out under controlled conditions in order to discover an unknown effect or law, to test or establish a hypothesis." Such experiments cannot be performed on the Earth system as a whole. Also, repetitions of experiments are unavailable, which normally help to rigorously sort out whether certain outcomes have emerged merely by chance or as a result of certain processes.

2. *The interaction of a myriad of processes in geophysical environmental systems.* One could argue that the same situation would prevail in a gas, with enormous numbers of molecules interacting with each other and responding to radiation. However, the temporal and spatial scales of the climate processes vary widely from, for example, the Hadley Cell in the tropical atmosphere (see Section 1.3.2) to turbulent eddies in the wake of a plane. Moreover, the dynamics at different scales are different in character and cannot be described by some universal similarity laws. Many of these processes exhibit chaotic behavior, with the overall effect that random-like variability (noise) emerges on all spatial and temporal scales (von Storch *et al.*, 2001). In principle, the system is deterministic but the many chaotic processes create a pattern of variability, which cannot be distinguished from the mathematical construct of random variations.

3. *The skill of models in describing the real world depends on the spatial scale* (e.g., Trigo and Palutikof, 1999; von Storch, 1995). Large-scale phenomena are usually better described than those occurring on smaller scales. Grid point values are

usually meaningful only if the variables are smooth (spatially homogeneous over some grid points) such that the grid point value is representative of a larger area. However, when the considered variables vary strongly from grid point to grid point, such as is usually the case, for instance, for precipitation, a correspondence between the values at a grid point and the values at the geographic location, formally corresponding to the grid point, is in most cases hardly possible. For larger areas, represented by many grid boxes, this no longer poses a problem.

Quasi-realistic models are used for several purposes (Müller and von Storch, 2004):

1. *Reconstructions.* When provided with adequate forcing, past marine climate variations, wave conditions, or storm surges may be reconstructed in some detail with the help of quasi-realistic models. When past decades are considered such reconstructions are useful to determine the tails of distributions (i.e., the likelihood of extreme events) and to analyze whether the statistics underwent any systematic change (e.g., in response to anthropogenic influences). Such reconstructions are needed when the observational evidence is either unavailable or corrupted by inhomogeneities; that is, factors not related to the quantity supposedly monitored (e.g., Jones, 1995; Karl *et al.*, 1993) (see Sections 4.2, 4.4).
2. *Scenarios.* Human action is changing the marine and coastal environment by global climate change related to the emission of greenhouse gases and aerosols into the atmosphere and by large-scale changes in land use. Also, on the more local scale, man is interfering with the hydrodynamics of the sea (e.g., by dredging shipping channels in estuaries), which may affect the height of tides and storm surges. The effect of such changes can be assessed by experiments conducted with quasi-realistic models (numerical experiments) in which the relevant conditions have been modified accordingly (see Section 4.6).
3. *Dynamical tests of hypotheses.* Such a hypothesis about the functioning of the system may come from observational evidence, theoretical speculations, or idealized process studies. With the help of numerical experiments the validity of such hypotheses may be tested. An example is the hypothesis that the transfer of momentum and energy from the atmospheric boundary layer to growing wind waves would have a significant effect on the emergence of storms (Doyle, 1995). This hypothesis was tested by a pair of year-long simulations with a regional atmospheric model (Weisse *et al.*, 2000). In one experiment, the drag of the sea surface on atmospheric flow was assumed to depend only on wind speed (Charnock, 1955), while in the second experiment a quasi-realistic wave model, which includes the effect of growing wind sea and the corresponding change in surface roughness, was coupled to the atmospheric model. Subsequently, the model output from both experiments was examined to see whether the presence of the dynamical wave model would have altered the intensity of storms. It turned out that the changes were small and did not point to any statistically significant difference between both experiments. Another example is the role of

82 **Models for the marine environment** [Ch. 3

the processes of tidal loading[2] and ocean self-attraction on the accuracy of modeling global ocean tides. Zahel (1978) tested this hypothesis and found that the inclusion of these two processes in quasi-realistic models significantly improves the simulation of tides in the world oceans. More examples are provided in Müller and von Storch (2004).

4. *Data analysis.* A more recent application concerns the analysis of environmental states. Because of the many degrees of freedom and practical barriers rendering certain variables unobservable,[3] a complete analysis of the state's system is impossible. However, intelligent use of the knowledge encoded in quasi-realistic models allows for a consistent interpretation of sparse and to some extent uncertain observations (e.g., Robinson *et al.*, 1998; Müller and von Storch, 2004). Such efforts are called data assimilation and their output is referred to as analysis (see Section 4.4).

5. *Forecasts.* A standard application of models is to prepare short-term forecasts. Based on some initial conditions the models are used to simulate some time into the future and the data derived from the model are used to make a prediction. In this book this application is not discussed any further. An overview of the approach can be found, for example, in Müller and von Storch (2004).

Before continuing with technical details and the validation of quasi-realistic models, we have to emphasize that even such quasi-realistic models are suffering from significant simplification of the complex real system even though they strive to be as realistic as possible. They can react in ways which cannot be foreseen by simple conceptual models. This is a virtue of such models as it makes them a kind of laboratory in which hypotheses can be tested. The models constitute a virtual or substitute reality (Müller and von Storch, 2004). In Chapter 5 we will present and discuss applications of quasi-realistic models to various reconstructions and scenarios of possible future changes of the marine climate. All of them are based on numerical experiments conducted with quasi-realistic climate, wind wave, and tide–surge models. In the following sections, these models are briefly discussed.

3.3 CLIMATE MODELS

There are many books and articles about climate models. The books by Washington and Parkinson (1986), McGuffie and Henderson-Sellers (1997), and von Storch *et al.* (1999) describe the challenges of numerical modeling on a technical level, while the monograph by Müller and von Storch (2004) deals more with the philosophical problems related to the use of such models. Also, the collection of papers offered

---

[2] Elastic deformation of the ocean bottom as a response to the moving tidal water bulge (see also Müller and von Storch, 2004).
[3] For instance, the spatial distributions of the transport of moisture in the atmosphere or of turbulent friction in the ocean.

by Trenberth (1993) or von Storch and Flöser (2001) and a description of the state of the art in the assessment reports provided by the Intergovernmental Panel on Climate Change (IPCC) (Houghton *et al.*, 1996, 2001; Solomon *et al.*, 2007) may be helpful for the interested reader.

Climate models are complex process-based dynamical models which operate on the entire globe or in limited regions of the world. Global climate models describe the dynamics of several compartments of the climate system; in particular, the atmosphere, the oceans, and sea ice. Most of them also describe, at least rudimentarily, surface hydrology, vegetation, or cycles of matter. Ice sheets and shelf ice are considered to be constant within the time periods simulated by contemporary climate models.

Atmospheric and oceanic dynamics are represented by so-called *general circulation models* (GCMs). GCMs of the atmosphere and the ocean are in most cases developed separately, with different codes and by different groups. The interactions between atmosphere and ocean are taken into account later by coupling these independently developed models. For these models, the not very informative term *coupled models* has frequently been used in the past.

When run for a few days, GCMs exploit the details of the initial state. The result of the simulation is a forecast with decreasing accuracy in time. When the same models are run for a longer period, the details of the initial state become irrelevant. Instead a *weather stream* is generated. If a different initial state is used in a second simulation, another weather stream is obtained, with different low-pressure systems, different details of the annual onsets of monsoons, etc. However, from both weather streams the same *statistics* are derived;[4] that is, both simulations return the same *climate* (see Section 1.2). In other words, the average position of mid-latitude storm tracks, the average annual number of cyclones, or the mean and standard deviation of the annual onsets of monsoons are the same within error bounds. As climate is thought to be conditioned by external forcing (see Section 1.4.1), such as the presence of greenhouse gases in the atmosphere, changing solar output, and other factors, changes in external forcing may cause changes in the statistics of weather. However, the weather itself varies independently of the presence of changing external factors.

3.3.1 General circulation models

General circulation models of the atmosphere and the ocean are based on differential equations describing the fluid dynamics of air, seawater, and sea ice. The differential equations used to describe oceanic and atmospheric dynamics are the *primitive equations*. These equations are named "primitive" not because they are simple, but because they are considered the most genuine equations describing atmospheric and oceanic dynamics. In pre-computer times, these equations could not be dealt with

---

[4] Within error bounds as a result of the limited integration time of the two simulations.

84 Models for the marine environment [Ch. 3

comprehensively and more heavily filtered approximations, like quasigeostrophic equations, were generally analyzed.

For the ocean, the primitive equations comprise the following state variables: velocity (currents) with horizontal components $\mathbf{u}_h = [u, v]$ and the vertical component w, density ρ, salinity S, temperature T, and pressure p. The primitive equations for the ocean read

$$\rho_0 \left(\frac{Du}{Dt} - \frac{uv}{a} \tan \varphi - fv \right) = -\frac{1}{a \cos \varphi} \frac{\partial p'}{\partial \lambda} + \mathcal{F}_u \tag{3.1}$$

$$\rho_0 \left(\frac{Dv}{Dt} + \frac{u^2}{a} \tan \varphi + fu \right) = -\frac{1}{a} \frac{\partial p'}{\partial \varphi} + \mathcal{F}_v \tag{3.2}$$

$$0 = -\frac{\partial p'}{\partial z} - g\rho' \tag{3.3}$$

$$0 = \frac{\partial w}{\partial z} + \nabla_h \cdot \mathbf{u}_h \tag{3.4}$$

$$\rho_0 \frac{DS}{Dt} = \mathcal{G}_S \tag{3.5}$$

$$\rho_0 \frac{DT}{Dt} = \mathcal{G}_T \tag{3.6}$$

$$\rho' = \rho'(S, T, p_o(z)). \tag{3.7}$$

with f being the Coriolis term; a the radius of Earth; and φ and λ representing latitudinal (north–south) and longitudinal (east–west) coordinates. The operators $\frac{D\cdot}{Dt}$, $\frac{\partial \cdot}{\partial \cdot}$, and $\nabla_h \cdot$ represent the material derivative, the partial derivative, and the horizontal divergence operator. $\mathcal{F}_u$, $\mathcal{F}_v$, $\mathcal{G}_S$, and $\mathcal{G}_T$ represent unresolved source and sink terms, such as surface stress or radiation. Density ρ is expressed as a sum of a reference density ρ_0 and a dynamically relevant deviation ρ' from that reference. Also, pressure is written as a reference p_0 plus a deviation p'.

From the seven state variables in (3.1)–(3.7), only four are *prognostic* variables (namely, horizontal velocities, salinity, and temperature) while the vertical velocity component, density, and pressure represent *diagnostic* variables. Here the words prognostic and diagnostic refer to the fact that the present state of a prognostic variable depends on the *past* state of the system's state variables, while those of diagnostic variables can be calculated solely from the present state. The vertical velocity component w is given by the divergence of horizontal flow (3.4), because of the balance of mass. From the four thermodynamic variables p, T, S, and ρ, two may always be expressed as functions of the other two. Here density and pressure are expressed as functions of S and T.

The primitive equations for the atmosphere are similar

$$\rho\left(\frac{Du}{Dt} - \frac{uv}{a}\tan\varphi - fv\right) = -\frac{1}{a\cos\varphi}\frac{\partial p}{\partial\lambda} + \mathcal{F}_u \tag{3.8}$$

$$\rho\left(\frac{Dv}{Dt} + \frac{u^2}{a}\tan\varphi + fu\right) = -\frac{1}{a}\frac{\partial p}{\partial\varphi} + \mathcal{F}_v \tag{3.9}$$

$$0 = -\frac{\partial p}{\partial z} - g\rho \tag{3.10}$$

$$\frac{D\rho}{Dt} + \rho(\nabla\cdot\mathbf{u}) = 0 \tag{3.11}$$

$$\rho\frac{Dq}{Dt} = \mathcal{G}_q \tag{3.12}$$

$$c_p\frac{DT}{Dt} = \mathcal{Q} + \alpha\frac{Dp}{Dt} \tag{3.13}$$

$$p = R_m(q)\rho T. \tag{3.14}$$

where the prognostic variables are the horizontal components of the wind u, v, temperature T, and specific humidity q, whereas the vertical wind component, pressure, and density represent diagnostic variables. Again, any two thermodynamic variables are considered to be prognostic and the other two diagnostic. Differently from the situation in the ocean, vertical movement is no longer considered to be given solely by horizontal divergence but also through the change of density (3.11). Also, the state variable salinity is replaced by specific humidity q. Note that for both the atmosphere and the ocean the vertical components of the equation of motion (3.3), (3.10) have been expressed in a simplified form; namely, the hydrostatic approximation, that is, $\frac{Dw}{Dt} = 0$.

The primitive equations need to be transformed into a discrete formulation before they can be solved numerically on a computer. Mainly, there are two different approaches to do so:

1. *The finite difference approach* replaces the differential operators in (3.1)–(3.13) by suitable discrete approximations and the solution is obtained iteratively by stepping forward with discrete time steps. Discretization is performed on more or less regular grids and, as a result, the model operates no longer with continuous variables but with discretized versions that represent averages for respective grid boxes. There are many different discretization algorithms, such as the leap-frog or the Runge–Kutta scheme (e.g., Atkinson, 1989) for discretization in time, or the various Arakawa grids for spatial discretization (Arakawa and Lamb, 1977).

86   **Models for the marine environment**   [Ch. 3

2.   *The Galerkin or spectral approach* is based on an orthonormal expansion of state variables and re-formulation of the equations in a new coordinate system, the spectral domain. The new state variables are then no longer, for instance, temperature at some location in three-dimensional space, but coefficients of a spherical harmonic of the temperature field (e.g., the global Wavenumber 3 coefficient). Compared with the finite difference approach, the spectral method is computationally more effective, but has some problems in keeping positive quantities such as humidity strictly positive.

The spectral approach is usually utilized in global atmosphere models, whereas regional atmosphere models, ocean, and wind wave models are mostly formulated using finite differences. The reason is that the spectral method requires a simple geometry so that a suitable, easy-to-handle system of orthonormal functions is available. In the case of the globe, these functions are the spherical harmonics. Ocean models describe irregularly formed regions, and regional atmospheric models are subject to complicated inhomogeneous boundary conditions. In the latter two cases simple orthonormal function systems are therefore not available. It is expected that the popularity of the spectral approach in atmospheric sciences will decrease with increasing available computer power.

The discretization necessary for numerical solution of the primitive equations leads to a loss of resolution in space and time. The selection of spatial resolution is a compromise between computational possibilities and the required length of integration. If the model is supposed to be integrated over 1,000 years,[5] then a coarser spatial resolution will be chosen than, under otherwise similar conditions, for an integration over a few days only. Note that the terms "resolution" and "grid size" are fundamentally different. Grid size is the length between any two grid points. As such it is a well-defined quantity. Resolution, on the other hand, describes the smallest scales properly described on the chosen grid. The latter is not a well-defined quantity (Pielke, Sr., 1991). Once the spatial grid size is set the time step cannot be determined independently. It is usually a function of spatial grid size and chosen such that numerical stability of the algorithm is ensured.

In the climate system, processes are operating at all space and time scales (see Section 1.3.1). The discretization required for numerical solution of the governing equations thus causes a cut-off of processes at a certain scale. Processes with time and spatial scales smaller than the resolution of the model cannot explicitly be accounted for in the integration. This truncation leads to many processes being disregarded; in particular, among the source and sink terms in (3.1)–(3.14). These processes are, however, essential for solution of the equations. For instance, cumulus convection can usually not be treated explicitly in GCMs. It is, however, essential for formation of the general circulation of the atmosphere. Such processes are therefore included in numerical equations in the form of *parameterizations*. In parameterizations, the

---

[5] For such an integration, spatial grid sizes of about 300 km are typically used these days. In order to achieve higher spatial resolution, regionalization or downscaling techniques have to be applied (see Section 4.5).

Sec. 3.3]   **Climate models**   87

expected net effect of unresolved processes on resolved ones is formulated as a function of resolved processes; that is, they are conditional upon large-scale (resolved) processes (see Section 1.4.2) and their overall effect can be determined or estimated, once the large-scale solution for the next time step is known. All models, atmospheric and oceanic, global and regional, contain many of these parameterizations, and they constitute a major cause for the different performances of different models under otherwise similar conditions.

3.3.2   Global climate models

The components of the climate system are sketched in Figure 1.1. In global climate models, these components are usually described in a more or less sophisticated form. Depending on the degree of sophistication and the period to simulate, the spatial grid size of such models may vary considerably. Figure 3.1 provides an impression of the spatial grid size of models presently in use. It illustrates how Europe is represented in these models. At T21[6] and T42 even the outlines of the European continent can hardly be deduced, while at T63 and T106 their representation becomes increasingly more realistic. Higher spatial resolution and more complex models need larger computing power. Therefore, presently only integrations of up to a length of several years are performed at T106 resolution while simulations extending over many decades, centuries, or millennia are done at resolutions of T42 and less.

When an ocean GCM and a sea ice model are coupled to an atmosphere model, the coupled model generally needs to be run for many decades, because otherwise the state of the different model components may still be inconsistent which results in artificial trends and variability that do no reflect internal climate variability but inconsistencies between the different model components. Thus, high-resolution atmosphere models (such as T106) are usually not coupled to an ocean model but are run with prescribed lower-boundary conditions such as sea surface temperatures or sea ice conditions. Depending on application the latter are usually obtained from observations or from simulations with fully coupled models at reduced spatial resolution. Such fully coupled models, which comprise not only an atmospheric but also an oceanic and other model components, presently have typical resolutions of T63 or less. Note that these numbers will change in the future when increasing computer power becomes available. The general statement that the higher complexity of models is paid with reduced spatial resolution (detail), however, probably remains.

Numerous aspects of the output of climate models have been compared with observations and analyses. In most cases—in particular, when large-scale hydrodynamic variables are considered—the similarity between simulated and observed (analyzed) states is good or at least reasonable. Small-scale thermodynamic variables (e.g., related to radiation and cloud processes) are less well represented. Figure 3.2

---

[6] This notation describes truncation (resolution) in the spectral domain (i.e., for spectral models). The letter (here T) denotes the type of truncation (here triangular) and the following number (here 21) the truncation wavenumber. For example, at T21 resolution zonal wavenumbers larger than 21 are cut off, corresponding to a spatial grid size of about $5.6° \times 5.6°$.

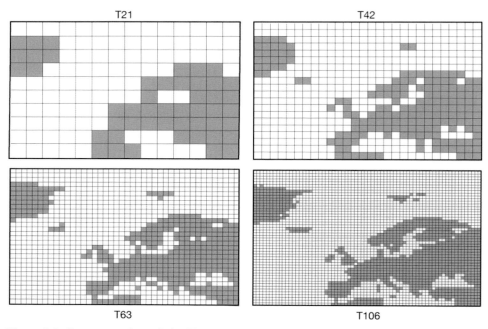

Figure 3.1. Representation of the European continent in the ECHAM global atmosphere model for different grid sizes ranging from T21 ($\approx 5.6°$) to T106 ($\approx 1°$). Courtesy of Ulrich Cubasch, Freie Universität, Berlin.

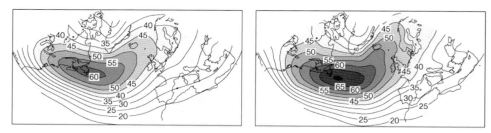

Figure 3.2. Location and intensity of the winter storm track over the North Atlantic as derived from a global climate model simulation at T42 resolution (left) and from operational weather analysis at the European Center for Medium Range Weather Forecast (ECMWF) at T106 resolution (right). In both panels the storm track is defined as the bandpass (2.5–6 days) filtered variance of 500 hPa geopotential height fields. Courtesy of Michaela Sickmöller.

provides an example of a large-scale variable; namely, the storm track (see Section 2.2) in the North Atlantic. In general, the spatial pattern and the location of the storm track are very similar when derived from a simulation with a global climate model or from operational weather analyses. The intensity of the storm track is somewhat smaller in the climate model simulation but the difference is usually considered acceptable within the range of other uncertainties.

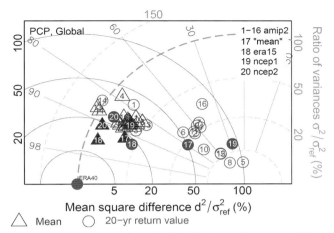

Figure 3.3. Quantitative comparison of spatial distributions of mean and 20-year return values of precipitation simulated by a series of climate models with ERA-40 reanalyses. Here "amip" (1–16) denotes the different GCMs that have been used in an inter-comparison study under the same conditions, "mean" (17) denotes the ensemble mean, and "era15", "ncep1", and "ncep2" (18–20) are the different reanalysis products. For details see text. Courtesy of Slava Kharin, University of Victoria. Data from Kharin et al. (2005).

Figure 3.3 provides an example of a more small-scale variable; namely, precipitation (Kharin et al., 2005). Here the performance of many different GCMs is compared with respect to the ERA-40 data set.[7] The performance of the GCMs was tested with respect to their ability to reproduce the observed spatial distribution of time-averaged precipitation and of 20-year return values. Three different statistics were computed for each GCM simulation: the globally averaged mean square difference between the ERA-40 reference and the model anomaly patterns normalized by the spatial variance of the reference pattern, the ratio of spatial variances of the reference and the model patterns, and the anomaly pattern correlation. These statistics are illustrated in Figure 3.3. The ERA-40 reference has a variance ratio of one, a mean squared difference of zero, and a pattern correlation of one. It can be found in the lower left margin of the plot. Ideally, the GCM simulations would have the same position or deviate only slightly from the ERA-40. However, the GCMs usually underestimate the spatial variance of simulated time averages by 50%–80% while the normalized mean squared differences are moderate (20–40%). The pattern correlation is generally high (typically about 80%). The spread for 20-year return values is much larger. Some of the models perform similarly well regarding mean precipitation, while other models produce a variability that is much too small (less than 20%) and large high mean square differences (up to 100%). Also, the pattern correlation is usually less than 80%, in some cases as little as 50% or less. Figure 3.3 additionally shows a comparison between different reanalysis products (ERA-40,

---

[7] A reanalysis (see Section 4.4) prepared by the European Center for Medium Range Weather Forecast (ECMWF).

90 Models for the marine environment [Ch. 3

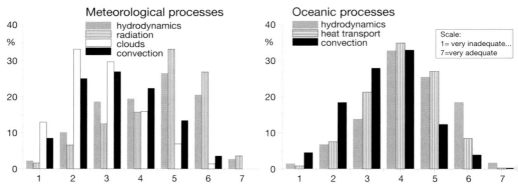

Figure 3.4. Results of a survey among climate modelers who were asked to subjectively assess the skill of atmospheric models (left) and ocean models (right). The modelers were specifically asked for their opinion about the quality of the representation of a number of different processes in the models (see insets). Courtesy of Dennis Bray, GKSS Research Center.

ERA-15, NCEP-1, NCEP-2). Again, substantial differences among the different products exist. This fact should be taken into account when analysis products are used to represent observed conditions in model comparison studies.

Apart from the fact that large-scale processes are generally better represented in global climate models than small-scale processes, there is no general statement about the quality or validity of such models. Instead, the answer depends on the problems, processes, and questions considered. However, an impression reflecting the opinion of climate modelers can be obtained. Bray and von Storch (2007) at the end of the 1990s conducted a survey among 190 climate modelers who were asked to assess the skill of contemporary climate models for a number of different processes (Figure 3.4). The scientists were requested to provide their assessment on a seven-graded scale ranging from "very inadequate" to "very adequate". It can be inferred that the assessment differs greatly for different processes. Hydrodynamics (i.e., implementation of the laws of conservation of mass and momentum) are generally believed to be adequately reproduced. Thermodynamic processes—such as those related to clouds, convection, or precipitation—are assessed by many experts as being represented insufficiently. While the survey may partly reflect the wish of modelers to continue their work in improving their models, the outcome of the survey also provides strong evidence that the models may indeed need some improvements. When the survey was repeated in 2003 only minor changes in this overall assessment could be inferred (Bray and von Storch, 2007).

3.3.3 Regional climate models

The main purpose of regional climate models[8] (RCMs) is to overcome the computational constraints of global climate models and to refine their spatial and

---
[8] Also referred to as limited area models (LAMs).

temporal resolution at least for limited areas. The main approach justifying this technique is to assume that the regional climate is to some extent controlled by or conditional upon the large-scale planetary climate (see Section 1.4.2). The latter is assumed to be represented well in global climate models (see Section 3.3.2). This assumption usually does not pose a problem in the case of analyses or reanalysis (see Section 4.4), but for climate simulations it represents a non-trivial approach. For instance, the formation of blocking situations constitutes large-scale information for many RCMs. In the global models this may, however, not be simulated well if their resolution is too low. The concept of refining GCM results with the use of regional GCMs (RCMs) is also referred to as *dynamical downscaling* (e.g., Giorgi and Mearns, 1991) (see Section 4.5).

Regional climate modeling is in most cases just regional atmospheric modeling with some rudimentary simulation of the thermodynamics of the upper soil layer. Other climatically relevant parameters at the Earth's surface—in particular, sea surface temperature, sea ice and lake ice conditions, or the state of the vegetation—are prescribed in most cases. In recent years, significant efforts have been made to construct coupled regional models that explicitly feature models of the regional oceans (e.g., shelf seas), lakes, runoff, or vegetation coupled with regional atmosphere models. Examples are the RCAO[9] model by the Swedish Rossby Center (Räisänen *et al.*, 2004) or the BALTIMOS[10] model system by the German Max-Planck-Institute for Meteorology (Jacob, pers. commun.). Both model systems comprise a Baltic Sea model, a regional hydrological model, and a regional atmosphere model.

In order to refine the results of global models, regional models need to be provided with information about the weather stream[11] generated by the global model. From these data, the RCMs construct regional weather streams that are consistent with both the weather stream generated by the global model and the physiographic details of the considered region. Presently, RCMs are typically run with spatial grid sizes between about 10 km and 50 km depending on the application and length of the simulation. For shorter integration periods finer grids of 3 km and less are possible. Figure 3.5 provides an impression of the resolution of such models. While regional details can hardly be inferred from contemporary global models, these are greatly enhanced in contemporary and future RCMs.

Commonly, regional models are forced by boundary conditions from global models. Usually these are provided every 6 hours, with some interpolation in-between. Boundary conditions are needed at the surface and along the lateral boundaries. Lateral conditions are enforced with the help of a *sponge zone* (Davies, 1976). Within the sponge zone, which consists of a few grid points along the lateral boundaries, the simulated state is nudged towards externally provided boundary conditions, with increasingly stronger nudging coefficients towards the RCM's margins. Technically, the term *nudging* refers to adding additional terms of the form

---

[9] Regional Coupled Atmosphere Ocean model.
[10] BALTIc MOdel System.
[11] That is, the sequence of synoptic weather situations.

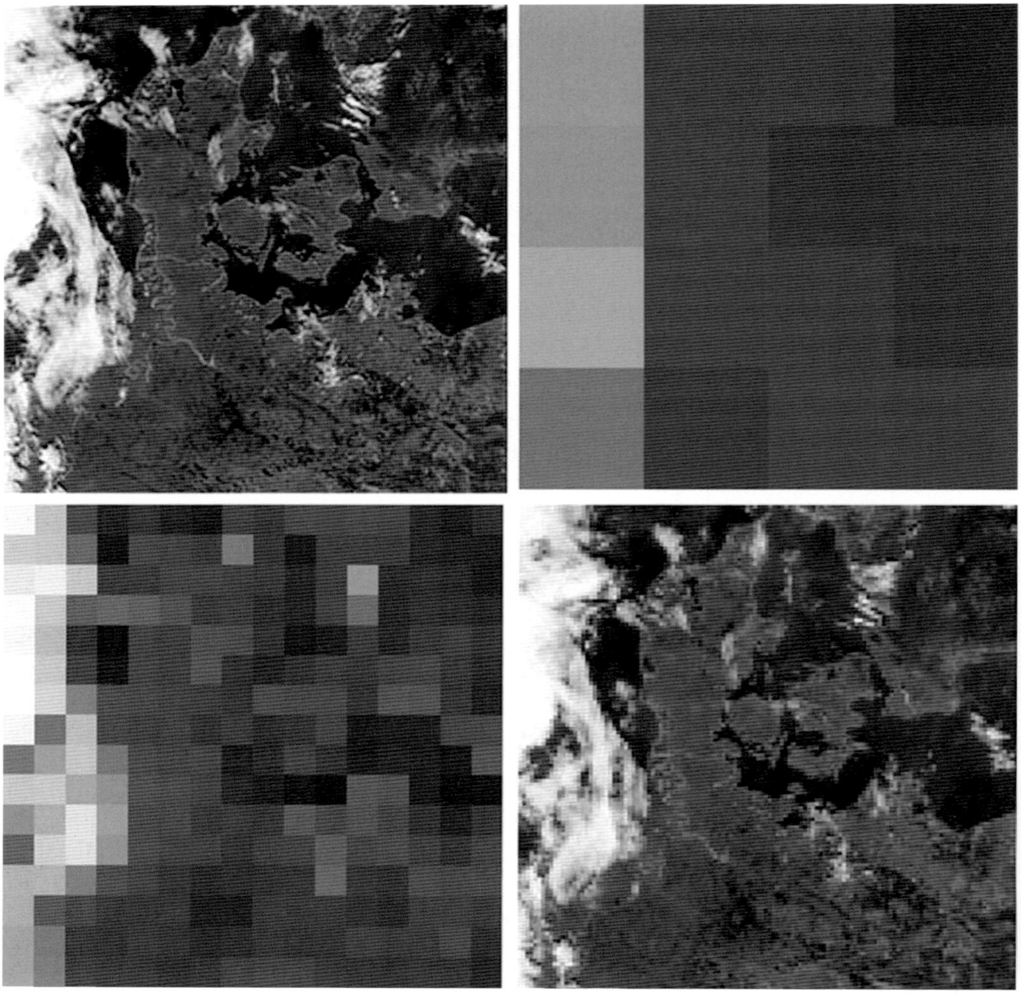

Figure 3.5. Gridded representation of Denmark and surrounding land and sea areas in a contemporary high-resolution global model (T63, about 200 km grid size, upper right), a contemporary regional model (about 50 km mesh size, lower left), and a regional model with 7 km mesh size, lower right). For comparison the corresponding satellite image is also shown (upper left). Courtesy of Burkhardt Rockel, GKSS Research Center.

$+\eta(\Psi_t^* - \Psi_t)$ to the dynamical equations, with η being a constant, Ψ_t^* the externally given time-dependent state to be enforced, and Ψ_t the dynamically determined time-dependent state. Practically, this nudging term forces the dynamical state Ψ_t of the regional model not to deviate too strongly from the prescribed state Ψ_t^* of the global model. In oceanography, this concept is sometimes also referred to as *Heney forcing*.

That this concept is practically working fine, at least in well-flushed[12] regions like northeastern America or northwestern Europe, has been convincingly demonstrated by Denis *et al.* (2002) in the so-called *Big Brother Experiment*. In this experiment an RCM was set up for a very large area using a 50 km grid. The model was then integrated over an extended period of time (Big Brother). Subsequently, a smaller domain within the larger area was chosen and boundary conditions for the smaller domain were extracted from Big Brother. An identical RCM was set up for the smaller domain using the same grid as in Big Brother (Little Brother). Little Brother was then forced with boundary data from Big Brother, but only after these were reduced to coarser resolutions (100, 200, and 500 km) typical of global GCMs. The question now was whether the fine-scale features simulated by Big Brother in the smaller domain (and which were removed from boundary forcing) would be recovered by Little Brother. Denis *et al.* (2002) found that this was indeed the case and that, after some spinup, the differences between the two simulations were small.

Mathematically, the problem of inferring the dynamical state of a fluid by providing lateral boundary conditions is not a well-posed problem. Lateral boundaries do not determine a unique solution in the interior of the model domain. Instead, several different states exist that are *all* consistent with a given set of lateral boundary conditions. The tendency to form different solutions in the interior as a response to the same boundary conditions depends on how well the region is flushed; that is, how efficiently boundary steering is established. In well-flushed areas this tendency is usually small, but it may become a big issue in areas with little through-flow like the Arctic. As a consequence, any two extended simulations that are run with identical boundary values but only slightly different initial states (which may simply be two observed states 12 hours apart) will show, more or less frequently, very different behavior. For well-flushed regions like Europe such divergent behavior is probably rare (Weisse *et al.*, 2000) while for less-flushed regions such as the Arctic it appears to more frequent (Rinke and Dethloff, 2000). This phenomenon of intermittent divergence reflects the competing influences of control by inflow boundary conditions and of regional chaotic dynamics.

Figure 3.6 shows an example of such intermittent divergence (Weisse and Feser, 2003). Here the observed near-surface wind speed at a location in the German Bight is shown together with corresponding simulated grid box values.[13] In this case, six simulated time series are shown (dashed). They are generated by the same regional

---

[12] A region is considered to be well flushed if disturbances entering the model domain at the inflow boundary are quickly advected through the model interior towards the outflow boundary. Whether or not a region is well flushed depends on the model setup and the area considered. Generally, regions comprising parts of extra-tropical storm tracks may be considered to be well-flushed while, for example, this is not the case for polar areas such the Arctic (Rinke and Dethloff, 2000).

[13] Note that there is a problem in comparing local wind measurements affected by local particularities with model-simulated grid box averages. Deviations between simulated and observed numbers may to some extent be due to local effects not described by the model's resolution.

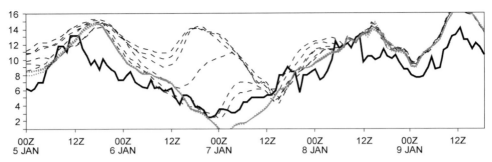

Figure 3.6. Example of intermittent divergence of solutions in a regional atmosphere model. Shown are the wind speeds at a small island off the German North Sea coast. The solid line displays the observed wind speed, the dashed lines the simulation in a series of equivalent but slightly differently initialized simulations in a conventional sponge zone setup. The gray band is obtained by running an ensemble from slightly different initial values with spectral nudging. Redrawn from Weisse and Feser (2003).

model, forced with the same (identical) lateral boundary conditions but with slightly different initial values. The reason for different developments in different simulations is not that the initial conditions are very different leading to different forecasts, but instead minuscule differences in the initial conditions excite chaotic behavior and divergence in the solutions of the RCM, all of which are consistent with prescribed boundary conditions. After a few days, here after January 8, boundary control prevails, divergence has ceased, and development is nearly the same in all six simulations. After several months, a similar episode emerges (not shown).

Spectral nudging. A method to overcome this intermittently emerging divergence of RCM solutions is to cast the regional modeling problem not as a boundary value problem but as a state space problem (e.g., Müller and von Storch, 2004). Here the regional model is used to augment existing knowledge about the regional state of the atmosphere derived from the regional model, with knowledge about the large-scale state of the atmosphere above a certain vertical level, where the influence of regional physiographic details is small. The latter is derived from the global model. In other words, provided there is some confidence in large-scale features simulated by the global model, additional constraints are applied, so that the regional model accepts only solutions that are not only consistent with the prescribed boundary conditions but also with the large-scale state of the global model over the entire regional model domain. The implementation of this concept leads to the *spectral nudging* approach (Waldron *et al.*, 1996; von Storch *et al.*, 2000; Miguez-Macho *et al.*, 2004) in which penalty terms are added in the equations of motion. These terms become large whenever the simulated large-scale state deviates from the prescribed large-scale state, but vanish when the regional model remains close to the prescribed large-scale state. An example of the effects of spectral nudging on RCM solutions is shown in Figure 3.6. Apart from the six simulations in which the conventional approach was used (dashed lines), six otherwise identical simulations using the

Table 3.1. Comparison of zonal and meridional wind speed variance at 850 hPa for different spatial scales. Shown are numbers from two RCM simulations with spectral nudging (nudging) and without spectral nudging (standard). Additionally, the same numbers are provided for the driving GCM (analysis). For details see von Storch *et al.* (2000).

Scale/variable	Units $(\mathrm{m}^2\,\mathrm{s}^{-2})$	Zonal wind			Meridional wind		
		Analysis	Standard	Nudging	Analysis	Standard	Nudging
Large scale	10^{-2}	1.6	1.2	1.6	1.4	1.3	1.5
Medium scale	10^{-6}	3.7	7.7	8.1	2.1	6.5	8.5

spectral nudging approach have been performed (gray lines). Compared with the conventional approach, spectral nudging not only effectively suppresses the emergence of intermittent divergence, but was also found to be better in capturing regional details (Weisse and Feser, 2003).

In the spectral nudging approach the coupling between the regional model and the large-scale state of the global model is performed in the spectral domain. First, the RCM solution is transformed into a Fourier representation from which the largest wavenumbers are nudged towards corresponding GCM wavenumbers. Subsequently the corrected large-scale state is transformed back into the RCM grid point space. This way, RCM variability at the largest scales is forced to be very similar to that prescribed by the driving GCM while variability at medium and small scales is not constrained. For wind fields this has been demonstrated by von Storch *et al.* (2000) who analyzed the variance of zonal and meridional wind speed components at 850 hPa obtained from two 3-month-long simulations with an RCM with and without spectral nudging applied. From their analyses two conclusions can be inferred (Table 3.1): first, variance at medium scales is similar in both the constrained and the unconstrained RCM simulation and variance is increased compared with the driving GCM (analysis). The latter may be interpreted as additional information provided by the RCM that is unresolved in the GCM data. Second, in the constrained simulation, variance at large scales is comparable with that obtained from the driving analysis while it is reduced for the unconstrained simulation. This implies improved performance of the constrained RCM simulation for those scales.

Added value of regional model simulations. A fundamental question that has been frequently ignored in the past is whether it is in principle possible to use RCMs to derive additional or more information about processes at unresolved scales in the driving GCMs. In other words, simulations with RCMs in general provide a spatially and temporally more detailed solution than obtained from the driving GCMs directly (see Table 3.1). However, the question remains on whether or not the additional information derived from the RCM simulations provides *added value*[14] or whether it just represents random fluctuations which do not enhance the

---

[14] That is, additional knowledge or skill.

96 **Models for the marine environment**   [Ch. 3

results derived from the driving GCMs directly. Unfortunately, this apparently simple question is not easy to answer. Moreover, there is no general answer. In particular it may depend on a number of different issues such as

1. *The specific problem or variable considered.* The added value may be different for, say, near-surface temperature or mean sea level pressure. The latter represents a field which has relatively little spatial variance on small scales while temperatures may vary strongly on much shorter distances. It may be more likely to add value for variables that show some signature on scales additionally resolved by the RCMs than for those that are mostly of large-scale character.
2. *The region considered.* Considering the same variable and/or problem, added value may vary from region to region. It may be expected that added value is larger when spatial variations in surface forcing are sufficiently strong over small distances. For example, surface roughness is relatively homogeneous over open water, while it varies strongly over land. For near-surface wind fields it may thus be expected that added value is enhanced in coastal areas when compared with open ocean conditions.

The question about the added value of RCM simulations has attained increasing attention in the last few years (e.g., Feser, 2006; Castro *et al.*, 2005). We will return to this question in more detail in Section 4.5.

3.4 WIND WAVE MODELS

Wave models are used to describe space and time variations of gravity waves at the sea surface (*sea state*). Here we focus on quasi-realistic wind wave or ocean wave models. In general, these models do not describe the variations or evolution of individual waves but of some statistical properties characterizing the sea state. In other words and using the terminology introduced in Section 2.4.2, these models are not designed to describe short-term variations within a given sea state, but to represent the longer variation between different sea states. For this purpose, it is more convenient to describe the sea state in the form of a *wave spectrum*. This concept is introduced in Section 3.4.1, which follows. The differential equation, which describes the dynamics of the wave spectrum, is presented and discussed in Section 3.4.2. Moreover, the different groups of wave models are briefly introduced. We conclude with a description of frequently used parameters that are derived from the spectrum and that are used to characterize the sea state (Section 3.4.3). Many more details about ocean waves and ocean wave modeling may be found, for example, in the textbook by Komen *et al.* (1994).

3.4.1 The wave spectrum

The sea surface is made up of waves on a broad range of scales, from millimeters to a few hundred meters. As a consequence, the height of the sea surface is highly variable, both in space and time. To start with, we note that a convenient way to describe the structure of a stationary and homogeneous[15] space and time–dependent field $\eta(\mathbf{x}, t)$ is to Fourier-decompose it into an infinite number of individual "waves"[16] with different amplitudes and wavelengths. For simplicity, we assume that η is centered; that is, the area mean water level has been subtracted

$$\eta(\mathbf{x}, t) = \int_{-\infty}^{\infty} \int_{-\infty}^{\infty} c_{\mathbf{k}}(t) e^{i\mathbf{k}\mathbf{x}} \, dk_x \, dk_y, \tag{3.15}$$

where $\mathbf{x}$ denotes location; $\mathbf{k} = [k_x, k_y]$ represents the two-dimensional wavenumber with wavelength $\lambda = 2\pi/|\mathbf{k}|$; and $c_{\mathbf{k}}(t)$ is the time-dependent complex amplitude of the wave $\mathbf{k}$. Introducing wave direction $\Phi_k = \mathbf{k}/|\mathbf{k}|$ as the direction of the wave-number vector, its components are given by $k_x = |\mathbf{k}| \cos(\Phi_k)$ and $k_y = |\mathbf{k}| \sin(\Phi_k)$. The three-dimensional wave spectrum $F_3(\mathbf{k}, \omega)$ is then given by

$$F_3(\mathbf{k}, \omega) = \int_{-\infty}^{\infty} c_{\mathbf{k}}(t) e^{i\omega t} \, dt. \tag{3.16}$$

The wave field at the sea surface is composed of *gravity waves* (see Section 2.4.1). Gravity waves satisfy the following dispersion relation between the angular frequency $\omega_{\mathbf{k}}$ and the wavenumber $\mathbf{k}$

$$\omega_{\mathbf{k}} = \pm\sqrt{g|\mathbf{k}| \tanh(|\mathbf{k}|h)} - \mathbf{k} \cdot \mathbf{u}, \tag{3.17}$$

where $\mathbf{u}$ is the speed at which the water body is moving relative to the ground (current velocity); h is the depth of the water body; g is the acceleration of gravity; and $\omega_{\mathbf{k}}$ denotes the frequency in s^{-1} (Hz). As a result of (3.17), the three-dimensional wave spectrum (3.16) can be reduced to two-dimensional spectra by integration and variable transformation. In particular, the following forms are frequently used

1. *Two-dimensional wavenumber spectra.* Two-dimensional wavenumber spectra are defined by

$$F_2(\mathbf{k}) = \int_{-\infty}^{\infty} F_3(\mathbf{k}, \omega) \, d\omega. \tag{3.18}$$

   The wavenumber spectrum is symmetric; that is, $F_2(\mathbf{k}) = F_2(-\mathbf{k})$ and its units are m^4.

2. *Two-dimensional wavenumber direction spectra.* Often the wavenumber spectrum is used as a two-dimensional spectrum of the scalar wavenumber $k = |\mathbf{k}|$ and the

---

[15] This means that the field has the same statistical properties in the entire domain, without "trends" in the mean or the level of variability. Ocean surface waves do have this property when the area considered is not too large. Note that the term "homogeneity", when used in the context of long time series, has a different meaning (see Section 4.2.1).

[16] Note that the term "wave" used here is not associated with gravity or any other type of physically defined waves. Instead, we are only dealing with the formal decomposition of a spatial field into two-dimensional trigonometric functions, colloquially also called "waves". Later in this section, we will explicitly make use of the perception that the elevation of the sea surface of the ocean is related to the presence of gravity waves.

98 **Models for the marine environment** [Ch. 3

direction $\Phi_k = \mathbf{k}/k$ or as a product of a one-dimensional wavenumber spectrum $F_1(k)$ and a directional spectrum $\Theta(k, \Phi)$ with $\int_\Phi \Theta(\mathbf{k}, \Phi)\, d\Phi = 1$.

3. *Two-dimensional frequency direction spectra.* The wavenumber spectrum may be rewritten and represented in the form of a frequency direction spectrum

$$F_2(\mathbf{k}) = \mathcal{F}_2(\omega_\mathbf{k}, \Phi_\mathbf{k}). \tag{3.19}$$

Since $\Phi_\mathbf{k}$ determines the direction, only the modulus of the frequency is relevant ($\Phi \in [0, 2\pi]$ and $\omega \geq 0$). The unit of $\mathcal{F}_2(\omega, \Phi)$ is m^2 s.

4. *One-dimensional spectra.* The frequency spectrum is frequently decomposed into

$$\mathcal{F}_2(\omega, \Phi) = \mathcal{F}_1(\omega)\Theta(\omega, \Phi) \tag{3.20}$$

with $\int_\Phi \Theta(\omega, \Phi)\, d\Phi = 1$. Here $\mathcal{F}_1$ is the one-dimensional frequency spectrum often also referred to simply as the *frequency spectrum*, and Θ is called the *directional spectrum*.

The various forms of wave spectra are also referred to as *variance spectra*, because they can also be understood as a statistical description of a randomly varying wave field, and can be derived by considering $\eta(\mathbf{x}, t)$ as a stationary random field with a spatial auto-covariance function $f(\xi, t) = \langle \eta(\mathbf{x}, t), \eta(\mathbf{x} + \xi, t) \rangle$ where $\langle \cdots \rangle$ denotes the cross-product. In this approach, the variance spectrum is then the Fourier transform of this auto-covariance function. Because wave energy is proportional to surface variance, wave spectra are also referred to as *power* or *energy spectra*.

Figure 3.7 shows an example of two different forms of a wave spectrum representing the same sea state. The spectra were derived from a nautical radar-based wave measurement system (WAMOS) (Nieto Borge *et al.*, 2004). The left panel shows the two-dimensional wavenumber spectrum. It can be inferred that the wavenumber spectrum is symmetric with energy concentrated at wavelengths around about 200 m. Since η is a real-valued function, the amplitudes appear in conjugate complex pairs; that is, $c_\mathbf{k} = c_{-\mathbf{k}}^*$. The amplitude $c_\mathbf{k}$ has a constant modulus $|c_\mathbf{k}|$ and a constantly changing phase $\pm\omega_\mathbf{k} t$. For positive angular frequencies, the wave is moving in the direction $\Phi = \mathbf{k}/|\mathbf{k}|$, for negative frequencies in the direction $\Phi_\mathbf{k} = -\mathbf{k}/|\mathbf{k}|$. Thus, each wavenumber is uniquely associated with a frequency $|\omega_\mathbf{k}|$ and a propagation direction $\Phi_\mathbf{k}$. Since the wavenumber spectrum is symmetric with respect to the origin ($F_2(\mathbf{k}) = F_2(-\mathbf{k})$), the wavenumber spectrum, as defined by (3.18), assigns variance jointly to both directions, $\pm\Phi_\mathbf{k}$. The right panel shows the same situation in the form of a two-dimensional frequency direction spectrum. Here the wave energy is concentrated around periods of about 11 s, indicating a swell system that propagates roughly towards the southwest.

3.4.2 Equation for wave energy

The basic equation for the description of a space–time varying sea state is a balance equation for what is called *wave action density* $N(\mathbf{k})$. Wave action density is

Sec. 3.4] Wind wave models 99

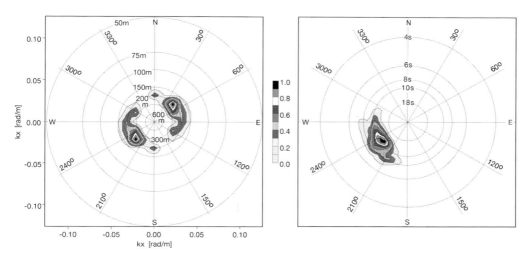

Figure 3.7. Two different forms of wave spectra representing the same sea state. Left: wavenumber spectrum in m^4. The x-axis and the y-axis represent the components of the wavenumber (k_x, k_y). For convenience, directions are given as well. Wavelength can be obtained from the distance to the center of the plot. Right: frequency direction spectrum in m^2 s (3.20). Here direction is given as in the left panel, and distance from the center denotes frequency. Both plots are derived from the same aerial measurement taken by a nautical radar-based wave measurement system (WAMOS) in the South Atlantic (Nieto Borge et al., 2004). For details about WAMOS see *http://www.oceanwaves.de* The diagram mostly shows swell with a wavelength of about 200 m and a period of about 11 s going roughly towards the southwest. Courtesy of Konstanze Reichert, OceanWaveS GmbH.

proportional to wave energy per frequency and is defined by

$$N(\mathbf{k}) = F_2(\mathbf{k})\sigma^{-1}(\mathbf{k}, h), \quad \text{with } \sigma = \sqrt{g|\mathbf{k}| \pm \tanh(|\mathbf{k}|h)}, \tag{3.21}$$

where σ denotes the so-called intrinsic frequency, representing the first term on the right-hand side of (3.17).

Introducing a four-dimensional vector $\mathbf{z} = [x_1, x_2, k_1, k_2]$ with x_1 and x_2 being spatial coordinates and k_1 and k_2 representing components of the wavenumber vector, the most fundamental form for wave action density balance reads

$$\frac{\partial}{\partial t}N + \frac{\partial}{\partial z_i}(\dot{z}_i N) = S_{in} + S_{ds} + S_{nl}, \tag{3.22}$$

where $\mathbf{c_g} = \dot{\mathbf{x}} = \dfrac{\partial \omega}{\partial \mathbf{k}}$ introduces the group velocity; while $\dot{\mathbf{k}} = -\dfrac{\partial \omega}{\partial \mathbf{x}}$ expresses conservation of the number of wave crests (Komen et al., 1994). The terms on the right-hand side of (3.22) denote so-called *source terms*; that is, the processes by which energy is transferred to and removed from the wave spectrum.

The first term on the right-hand side of (3.22) S_{in} denotes the energy transfer from the wind into the wave field (*wind input*). For the time scales considered here, wind represents the only source of energy for the generation of sea surface waves. The

100 Models for the marine environment [Ch. 3

transfer of energy is roughly proportional to the square of wind speed. Therefore, the quality of the wind fields used to drive numerical wave models is of crucial importance for the quality of wave simulations. There are generally two different processes involved: initially when the sea surface is still smooth and calm, small turbulent atmospheric pressure variations are responsible for inducing perturbations in sea surface height (Phillips, 1957). Once the perturbations have reached sufficient size, the air flow above the growing waves is disturbed and shear flow instability develops (Miles, 1957) (see also Section 2.4.1). Wind input is generally parameterized as being proportional to the energy of the waves

$$S_{in} = \gamma N, \tag{3.23}$$

indicating that the extraction of momentum from the atmosphere to the wave field is stronger when waves are taller. The growth rate γ is determined by the *friction velocity* u_* the square of which is proportional to the magnitude of the wind stress $\tau = u_*^2 \rho_a$ where ρ_a denotes air density. Friction velocity is related to the wind speed $u^2 = u_*^2 c_d^{-1}$ by the drag coefficient c_d which varies with wind speed and the sea state itself (Janssen, 1991; Makin *et al.*, 1995).

The second term on the right-hand side of (3.22) S_{ds} denotes the *dissipation* of wave energy from the spectrum. In principle, wave energy can be dissipated by three different processes:

1. *Whitecapping* is the primary dissipation process in open oceans and under deep-water conditions. It occurs when the waves grow and, as a consequence, become increasingly steeper. Once a critical point is reached, the waves break. The process thus depends on wave energy and steepness. It limits wave growth in open seas and transfers energy to underlying currents. Whitecapping is observed in the form of *spilling* breakers.
2. *Surf breaking* is relevant only in extremely shallow waters where wave height and water depth are of comparable orders of magnitude. When a wave propagates into extremely shallow water the upper part of the wave becomes faster than the lower part. At some point the crest overtakes the preceding trough and surf breaking occurs. This is referred to as *plunging* breakers. An example is shown in Figure 2.16.
3. *Wave-bottom interaction* refers to the dissipation of wave energy by interaction of the waves with the bottom. Outside the surf zone and away from deep waters, wave-bottom interaction is usually the most relevant dissipation process. Wave-bottom interaction may take different forms which include bottom friction, percolation (water penetrating into the bottom), or the movement of seabed material.

The last term on the right-hand side of (3.22) S_{nl} represents weakly nonlinear wave–wave interactions (Hasselmann and Hasselmann, 1985). These processes do not change the total energy of the wave field but redistribute energy within the spectrum. Their most important effect is to shift energy from higher to lower fre-

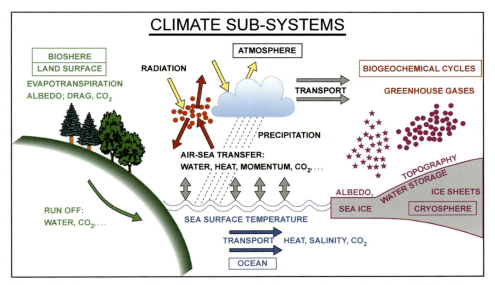

Figure 1.1. Components of the climate system. Redrawn after von Storch and Hasselmann (1995).

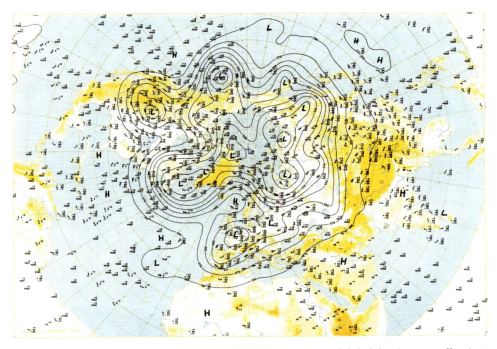

Figure 2.4. Circumpolar representation of 500 hPa geopotential height (contour lines) at 12:00 UTC, October 5, 1998. Reproduced from the *European Meteorological Bulletin* published by and with permission of the German Weather Service; courtesy of Elke Roßkamp, German Weather Service.

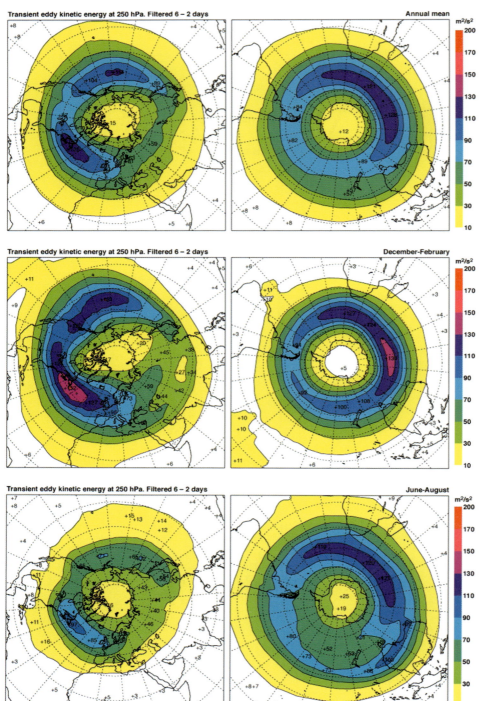

Figure 2.6. Storm track as represented by 2-day to 6-day eddy kinetic energy at 250 hPa from ECMWF ERA-40 reanalysis. From top to bottom: annual mean, December–February mean, June–August mean—Northern Hemisphere (left), Southern Hemisphere (right). Redrawn after Kallberg et al. (2005).

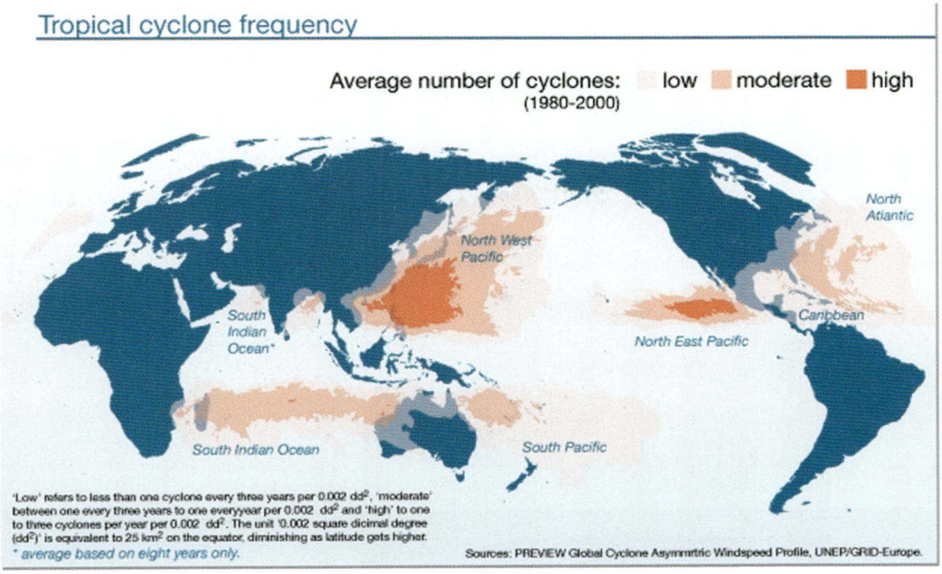

Figure 2.8. Tropical cyclone frequency. For details see graphics. From UNEP/GRID-Arendal, tropical cyclone frequency, UNEP/GRID-Arendal Maps and Graphics Library, http://maps.grida.no/go/graphic/tropical-cyclone-frequency

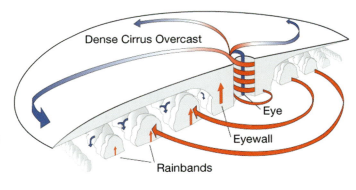

Figure 2.9. Schematic sketch of tropical cyclone structure. Arrows indicate principal air flow of relatively warm (red) and cold (blue) air.

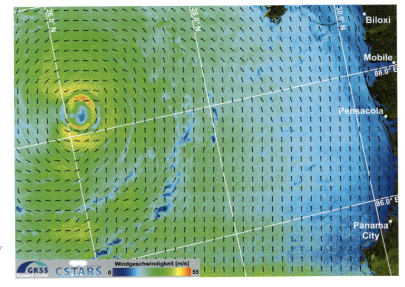

Figure 2.11. Wind field of Hurricane Katrina at 15:50 UTC, August 28, 2005 prior to making landfall at the U.S. coast. The wind field was derived from the Advanced Synthetic Aperture Radar (ASAR) aboard the European satellite ENVISAT using the WiSAR algorithm of Horstmann *et al.* (2005). Courtesy of Wolfgang Koch, GKSS Research Center Geesthacht.

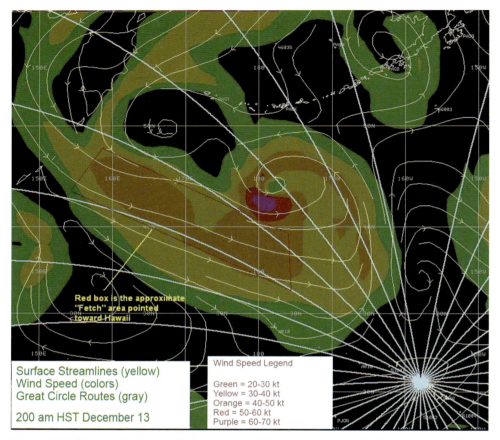

Figure 2.15. Meteorological conditions responsible for extreme surf heights at North Shore, Oahu (Hawaii) on December 15, 2004. The figure shows a low-pressure system over the northwest Pacific with corresponding strong wind fields (colors). The white lines denote the great circle lines along which swell may propagate towards Hawaii. The red box marks the high-wind speed areas (fetch) in which the wave system was generated. From *http://www.prh.noaa.gov/hnl/pages/events/dec04surf/Extreme_North_Shore_Surf.php*, reproduced with permission from NOAA.

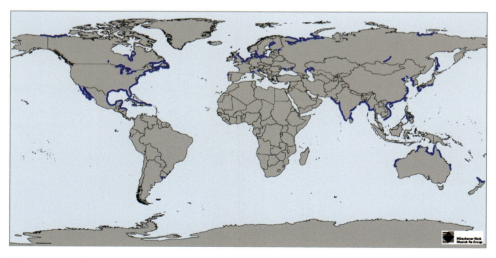

Figure 2.26. Global map of areas where storm surges pose a significant threat to the environment (blue). Courtesy of Peter Höppe, Munich Re.

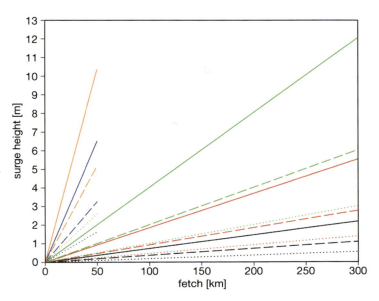

Figure 2.25. Dependence of surge heights on wind speed, fetch, and water depth. The curves are shown for 10 m (solid), 20 m (dashed), and 40 m (dotted) water depth, and for wind speeds of 8 (black), 10 (red) and 12 (green) Beaufort wind speeds, representing weak, moderate, and strong extra-tropical storms, respectively. Curves for Category-3 (blue) and Category-4 (orange) hurricanes are shown in addition. Because of the smaller spatial scale of hurricanes when compared with extra-tropical storms, the curves for tropical storms are shown only up to 50 km fetch.

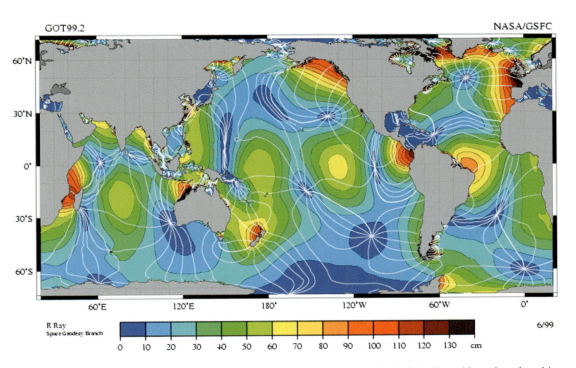

Figure 2.29. Principal lunar (semidiurnal) tide in the world oceans. Amplitude is indicated by color, the white lines are cotidal lines differing by 1 h. Data from Topex/Poseidon, courtesy of Richard Ray, NASA.

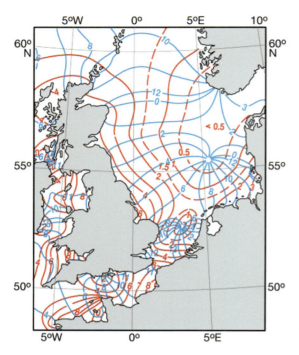

Figure 2.30. Tides in the North Sea. Cotidal lines of the principal lunar (M2) tide in hours after the Moon's transit through the Greenwich Meridian (blue) and mean tidal range in meters at spring tide as the sum of M2 and S2 (principal solar) tidal constituents. Redrawn after and courtesy of Matthias Tomczak, Flinders University.

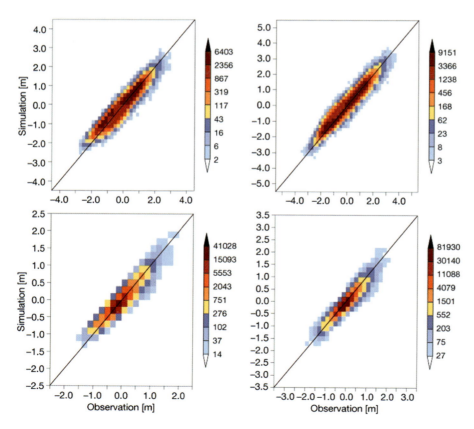

Figure 3.9. Scatter plot between observed (x-axis) and simulated (y-axis) hourly water levels (top) and surge levels (bottom) at Borkum (left) and Cuxhaven (right), two tide gauges in the German Bight. Borkum is more representative for open sea conditions while Cuxhaven is located in an estuary. Storm surges were extracted from total water levels by harmonic analysis taking 30 major tidal constituents into account. Colors indicate the numbers of cases in each 0.2×0.2 m box. Redrawn after Weisse and Pluess (2006).

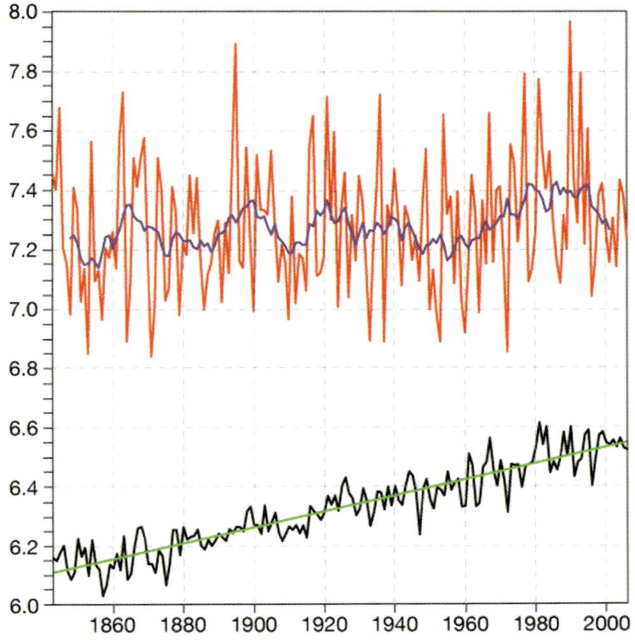

Figure 4.12. Annual mean high-water levels and linear trend at Cuxhaven, Germany (bottom) and the corresponding difference between annual 99-percentile and annual mean high-water levels (top); In addition, an 11-year running mean is shown in the upper panel. Redrawn and updated after von Storch and Reichardt (1977).

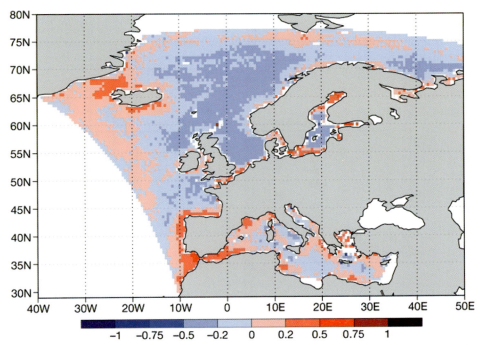

Figure 4.17. Skill score S (4.1) between collocated Quickscat wind speeds, NCEP reanalysis, and dynamically downscaled NCEP reanalysis wind speeds. Data were collocated for 2000–2007 for Quickscat wind speeds smaller than 25 m s^{-1}. Positive values (red) indicate an improvement, negative values (blue) a decline of regionally downscaled wind speeds in comparison with NCEP reanalysis wind speeds, when Quickscat data are considered as "true". Redrawn after Winterfeldt and Weisse (2009). Courtesy of Jörg Winterfeldt, GKSS Research Center.

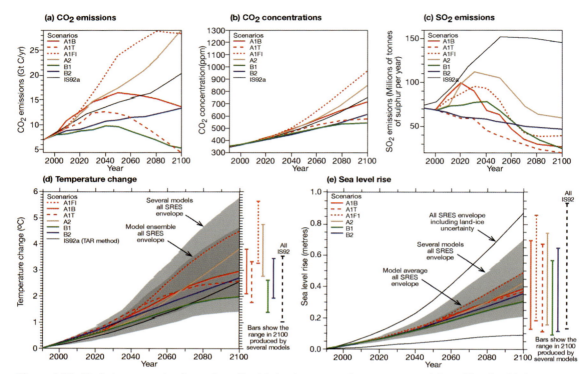

Figure 4.18. Emission scenarios for carbon dioxide in gigatons carbon per year and for sulfur dioxide in megatons sulfur per year for different SRES scenarios. Additionally, corresponding carbon dioxide concentrations and climate change projections for global mean temperature and sea level are shown. Redrawn from Houghton *et al.* (2001). Reproduced by permission of the IPCC/Cambridge University Press.

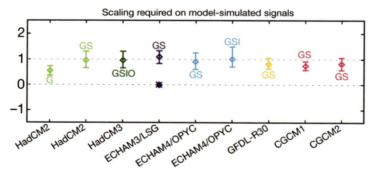

Figure 4.19. Estimate of scaling factors by which the response of different models to different forcing functions has to be multiplied to be in agreement with observed changes in global mean temperature. Vertical bars indicate the 5–95% uncertainty range due to internal variability. The names on the *x*-axis denote different climate models, the capital letters in the figure the different forcings; namely, greenhouse gases (G), greenhouse gases and sulfate (GS), greenhouse gases, sulfate, and indirect sulfate forcing (GSI), and greenhouse gases, sulfate, indirect sulfate, and stratospheric ozone forcing (GSIO). For further details see Mitchell *et al.* (2001). Redrawn from Mitchell *et al.* (2001, Fig. 12.12). Reproduced by permission of the IPCC/Cambridge University Press.

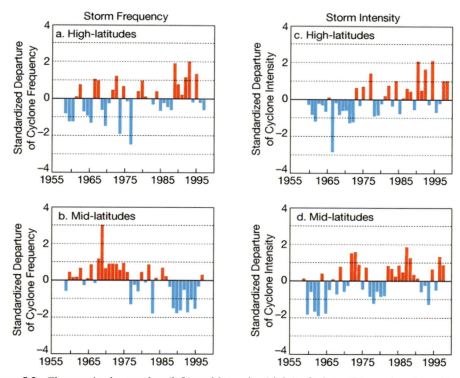

Figure 5.2. Changes in the number (left) and intensity (right) of winter (November–March) extra-tropical cyclones in the Northern Hemisphere in high (60–90°N, top) and mid-latitudes (30–60°N, bottom) 1959–1997. Data are shown as anomalies normalized by standard deviation. For details see McCabe *et al.* (2001). Redrawn after McCabe *et al.* (2001). Reproduced/modified by permission of American Meteorological Society.

Figure 5.3. Storm index for northwest Europe based on geostrophic wind speed percentiles according to the methodology described in Alexandersson *et al.* (1998). Blue circles are 95-percentiles and red crosses 99-percentiles of standardized geostrophic wind speed anomalies averaged over 10 sets of station triangles. The blue and red curves represent a decadal running mean. Redrawn from Trenberth *et al.* (2007, Fig. 3.41). Reproduced by permission of IPCC/Cambridge University Press.

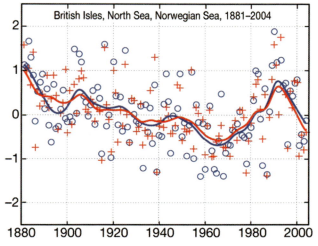

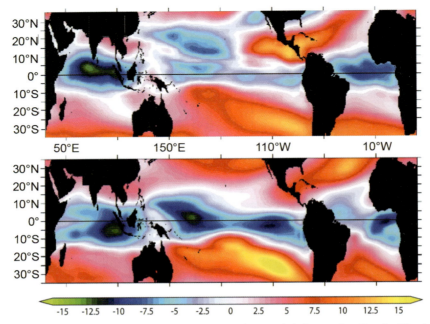

Figure 5.5. Change in vertical wind shear in percent per degree global warming for the Northern Hemisphere (top) and the Southern Hemisphere (bottom) tropical cyclone season. Redrawn from Vecchi and Soden (2007). Published/copyright (2007) American Geophysical Union. Reproduced/modified by permission of American Geophysical Union.

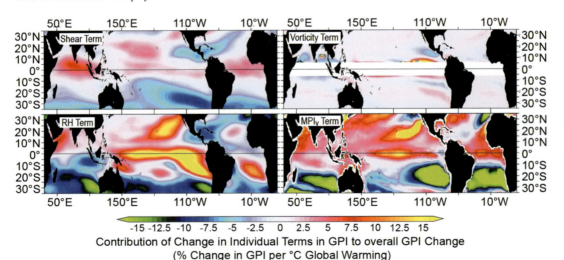

Figure 5.6. Ensemble mean of the contribution of change in individual terms in the genesis potential index (GPI)—Equation (2.5)—to overall GPI change in percent. Redrawn from Vecchi and Soden (2007). Published/copyright (2007) American Geophysical Union. Reproduced/modified by permission of American Geophysical Union.

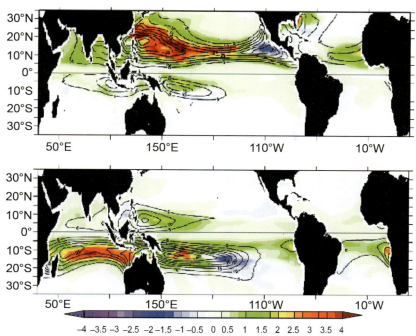

Figure 5.7. Change in genesis potential index (GPI)—see Equation (2.5)—per degree global warming for the Northern Hemisphere (top) and the Southern Hemisphere (bottom) tropical cyclone season. Redrawn from Vecchi and Soden (2007). Published/copyright (2007) American Geophysical Union. Reproduced/modified by permission of American Geophysical Union.

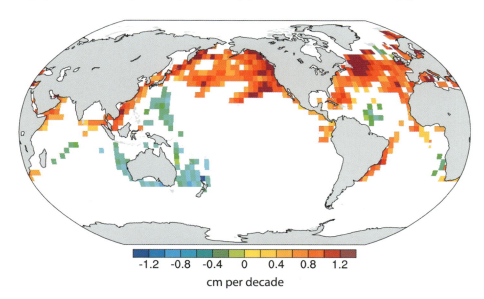

Figure 5.8. Estimates of linear trends 1950–2002 in significant wave height in centimeters per decade derived from VOS data. Trends are only shown for locations where they are statistically significant at the 5% level. Redrawn from Trenberth *et al.* (2007, Fig. 3.25) as adapted from Gulev and Grigorieva (2004). Reproduced by permission of the IPCC/Cambridge University Press.

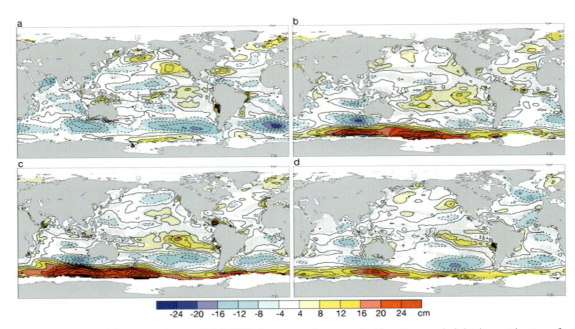

Figure 5.9. Ensemble mean change 2080–1990 in seasonal mean significant wave height in centimeters for January, February, March (a), for April, May, June (b), for July, August, September (c), and for October, November, December (d) for the A2 emission scenario. Areas where changes are statistically significant are hatched. Redrawn from Wang and Swail (2006a). Courtesy of Xiaolan L. Wang, Environment Canada.

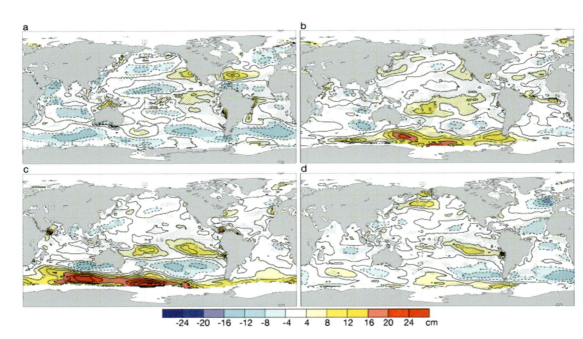

Figure 5.10. Same as Figure 5.9 but for the B2 scenario. Redrawn from Wang and Swail (2006a). Courtesy of Xiaolan L. Wang, Environment Canada.

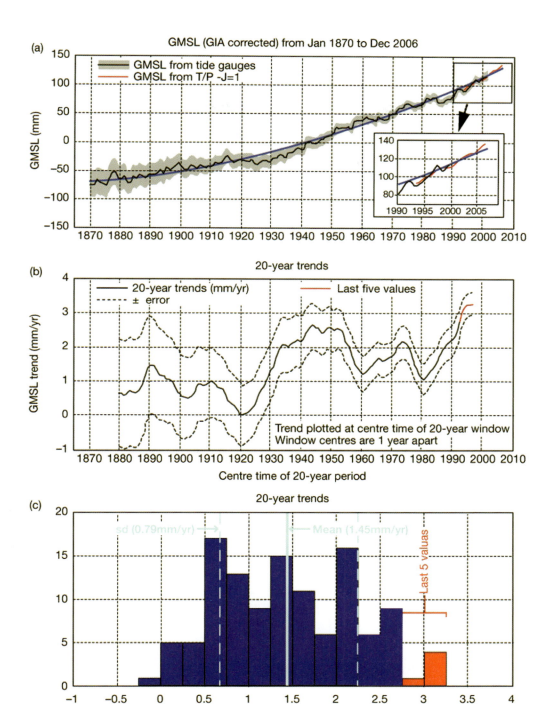

Figure 5.12. Global mean sea level 1870–2006 with error estimates (gray) (a, top); linear trends over 20-year periods with error estimates (dotted) (b, middle) and corresponding histogram (c, bottom). Redrawn from Church *et al.* (2008). Courtesy of John Church, CSIRO, Australia.

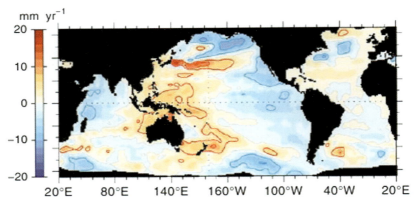

Figure 5.13. Regional rates of sea level rise 1993–2003 in millimeters per year obtained from satellite altimetry and plotted around the global mean value for the same period. Redrawn from Church *et al.* (2008). Courtesy of John Church, CSIRO, Australia.

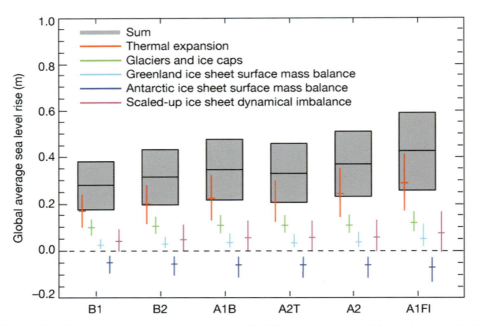

Figure 5.14. Projections of and uncertainty in (5–95% ranges obtained from the spread of model results) global mean sea level rise and its components between 2090–2099 and 1980–1999 for different emission scenarios. The projected total (gray) does not include the contribution from the scaled-up ice sheet response (magenta). Redrawn from Meehl *et al.* (2007, Fig. 10.33). Reproduced by permission of IPCC/Cambridge University Press.

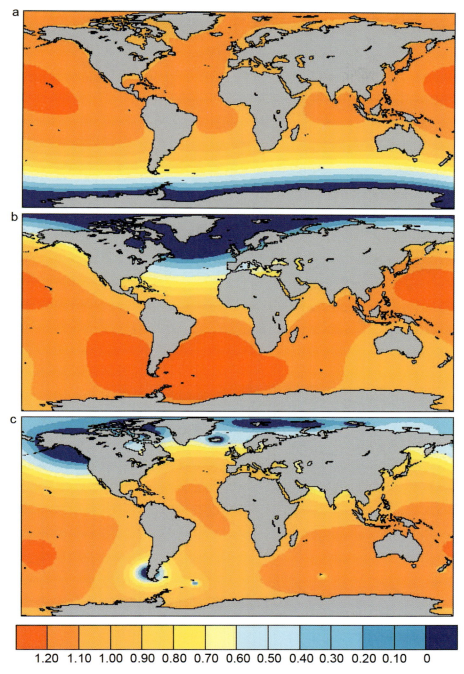

Figure 5.15. Fingerprints of relative sea level change due to ice mass variations for Antarctica (a, top), Greenland (b, middle), and mountain glaciers and ice sheets (c, bottom). Reprinted/adapted by permission from Macmillan Publishers Ltd.: *Nature*, Mitrovica *et al.* (2001), copyright 2001.

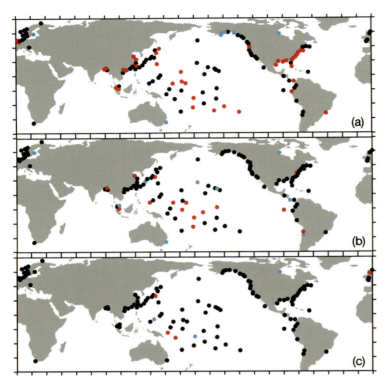

Figure 5.16. The distribution of t[he] gauge stations in the analysis of Woodworth and Blackman (2004[).] Stations with observed trends in annual 99-percentile water levels significantly different from zero a[re] shown in red (positive trend) or blue (negative trend) while others are shown in black (top); as befo[re] but with 99-percentile time series reduced by annual medians (middle); as before but with 99-percentile time series reduced by annual medians and with the tid[al] contributions to the percentiles removed (bottom). Redrawn fro[m] Woodworth and Blackman (2004[).] Reproduced/modified by permiss[ion] of American Meteorological Society.

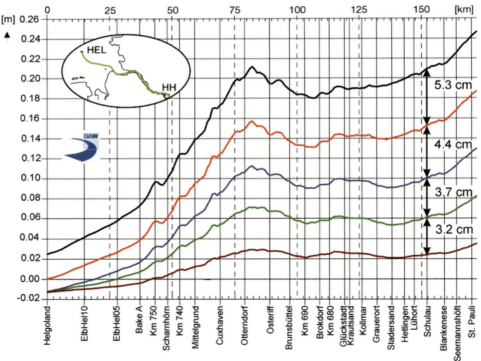

Figure 5.17. Changes in mean tidal range caused by a mean sea level rise of 20 cm (brown), 40 cm (green), 6[0 cm] (blue), 80 cm (red), and 100 cm (black) along a cross-section from Helgoland (HEL), the central island i[n] German Bight, to Hamburg (HH) upstream in the River Elbe. The inset indicates the location of the cross-sec[tion.] Redrawn after Pluess (2006) and courtesy of Andreas Pluess, Federal Waterways Engineering and Rese[arch] Institute, Hamburg, Germany.

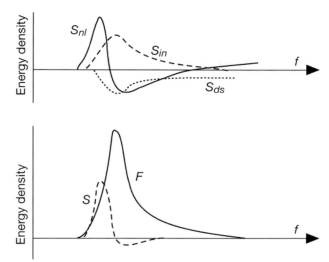

Figure 3.8. Principal balance of wind input (S_{in}), dissipation (S_{ds}), and nonlinear wave–wave interactions (S_{nl}) (top) and of wave energy and the overall contribution from different source terms (bottom).

quencies. As a consequence, the waves dominating a sea state become increasingly longer with the passage of time from the onset of the wind field.

Figure 3.8 illustrates the principal balance of source terms relative to frequency. Near the spectral peak, energy input from the wind field usually exceeds dissipation. The excess of energy is then transferred to higher and lower frequencies by nonlinear wave–wave interactions. At higher frequencies, energy is dissipated while the energy transfer to lower frequencies leads to a shift of the spectral peak over time; that is, as the wind continues to blow the dominating wave components usually become longer. In a growing wind sea the overall sum over all source terms is generally positive and wave energy (and thus wave height) increases. At some stage in the process the wind input is eventually balanced by dissipation. Subsequently, the wave field no longer grows and a sea state referred to as a *fully developed* sea state is reached.

Existing quasi-realistic wind wave models differ mainly in the way they handle the wave energy equation (3.22). So-called first-generation and second-generation models basically operate with equations for parameters describing the shape of the wave energy spectrum while third-generation models solve (3.22) explicitly. Presently, third-generation models are widely used while second-generation models may still be applied in some situations, especially when computational resources are limited.

Third-generation models may differ in the way different source terms are formulated and/or how they are represented or neglected in these models. Two of the most widely distributed models in this group are the wave model WAM (WAMDI-Group, 1988) and the WAVEWATCH model (Tolman, 1991). Both models are constructed for integration over horizontally extended areas, so that the life time of a wave field is sufficiently long for nonlinear interactions to play a significant role while wave-bottom interactions are less relevant. As such these models are mostly used for open ocean and shelf sea conditions. For coastal applications, weakly nonlinear interactions are less relevant but wave-bottom interaction becomes

102   **Models for the marine environment**   [Ch. 3

increasingly important. Third-generation models with a special focus on coastal and shallow-water conditions comprise the K-model (Schneggenburger *et al.*, 1997) or the SWAN model (Booij *et al.*, 1999). Such models may also include additional terms that are relevant only in extremely shallow water or may ignore some deep-water terms.

3.4.3   Frequently used parameters to describe the sea state

Ocean wave models generally describe the evolution of the wave spectrum (3.22) in one or other form. Mostly the interest is, however, not in the spectrum itself but in some parameters such as wave height or wave period. These parameters can be derived from the spectrum and are usually provided as additional model output. In the following we describe some of the most frequently used parameters.

The parameter most frequently used to describe the sea state is *significant wave height*. Historically, it was defined the following way: from a time series of wave height recordings taken during the prevalence of one particular sea state the largest third is selected. The average of these largest waves is then referred to as the significant wave height $H_{1/3}$ that characterizes that particular sea state. It corresponds to the average height of the highest waves in a wave group. Thus, significant wave height is usually rather similar to the wave height an experienced observer would report as the "prevailing" wave height by visual observations. The term was originally coined by Walter Munk during World War II, when first attempts for wave forecasting were developed in preparation of U.S. landing operations and where the term "significant wave height" corresponded to the wave height perceived by the drivers of the landing crafts.

Significant wave height can also be determined from the wave spectrum. To do so, the spectrum is usually expressed in terms of the moments of the spectrum (distribution) where the nth-order moment m_n of the spectrum is defined by

$$m_n = \int_0^\infty \int_0^\infty \omega^n \mathcal{F}_2(\omega, \Phi)\, d\omega\, d\Phi. \tag{3.24}$$

In this definition the zero-order moment m_0 represents the variance of the wave field. It is therefore used for definition of wave height parameters derived from the spectrum. It can be shown that a wave height parameter corresponding as closely as possible to the significant wave height $H_{1/3}$ derived from a wave record can be obtained by

$$H_{m_0} = 4\sqrt{\int_{\omega,\Phi} \mathcal{F}_2(\omega, \Phi)\, d\omega\, d\Phi} = 4\sqrt{m_0}. \tag{3.25}$$

In theory the correspondence between H_{m_0} and $H_{1/3}$ is valid only for very narrow spectra, but in most cases the difference is in the order of 5% only (WMO, 1998). As both definitions of significant wave height lead to slightly different results, the significant wave height determined from the spectrum is generally referred to as H_{m_0} to distinguish it from $H_{1/3}$ as derived from a wave record.

Sec. 3.5]

The following parameters are also frequently used: the *peak wave frequency* f_p is the frequency that corresponds to the peak of the spectrum; that is, the frequency at which $\int \mathcal{F}_2(\omega, \Phi)\, d\Phi = \max$. The *peak period* T_p is the period corresponding to f_p. The T_{m01} *period* is defined by $T_{m01} = \dfrac{m_0}{m_1}$ and represents the wave period corresponding to the mean frequency of the spectrum. The T_{m02} *period* is defined by $T_{m02} = \sqrt{\dfrac{m_0}{m_2}}$. Theoretically, it is equivalent to the mean zero-downcrossing period[17] obtained from a wave record. The T_{m02} period is, however, sensitive to the high-frequency cut-off in the integration of (3.2.4). For buoy data, this cut-off typically occurs at about 0.5 Hz. For wave models, a high frequency tail is often fitted to the modeled spectrum before the moments are computed.

3.5 TIDE–SURGE MODELS

Numerical tide–surge models are based on integration of a simplified set of equations (3.1)–(3.7) used to integrate the oceanic component of a climate model. In particular, the models are integrated without thermodynamic contributions and geometric terms $\dfrac{uv}{a}\tan(\varphi)$ and $\dfrac{u^2}{a}\tan(\varphi)$, which become irrelevant in regional applications. When the forcing terms $\mathcal{F}$ are parameterized as horizontal and vertical diffusion $K_h \nabla_h^2 \mathbf{u}$ and $\dfrac{\partial}{\partial z}\left(K_v \dfrac{\partial \mathbf{u}}{\partial z}\right)$ the relevant equations for a tide-surge model read

$$\frac{\partial \mathbf{u}_h}{\partial t} + \mathbf{u}_h \cdot \nabla \mathbf{u}_h + f\mathbf{k} \times \mathbf{u}_h - K_h \nabla_h^2 \mathbf{u}_h - \frac{\partial}{\partial z}\left(K_v \frac{\partial \mathbf{u}_h}{\partial z}\right) = -\frac{1}{\rho}\nabla_h p \qquad (3.26)$$

$$\frac{\partial p}{\partial z} = -\rho g \qquad (3.27)$$

$$\nabla_h \cdot \mathbf{u} = 0, \qquad (3.28)$$

where (3.26) represents the horizontal components of Navier–Stokes equations; (3.27) represents the hydrostatic approximation[18] that replaces the vertical component of the Navier–Stokes equation; and (3.28) is the continuity equation.

The Navier–Stokes equation (3.26) is the equation of motion for the two components $[u, v]$ of the horizontal current $\mathbf{u}_h$. The unit vector $\mathbf{k}$ points in the vertical direction, f is the Coriolis parameter, K_h and K_v are horizontal and vertical eddy viscosity coefficients, and p denotes the pressure field. Decomposing (3.26) into

---

[17] After the average water level has been determined it is possible to identify individual waves in the wave record. The average time between individual downward crossings of the mean water level line is referred to as the zero-downcrossing period.

[18] For the dynamical background of this approximation see Müller and von Storch (2004).

104 **Models for the marine environment** [Ch. 3

velocity components u and v yields

$$\frac{\partial u}{\partial t} + u\frac{\partial u}{\partial x} + v\frac{\partial u}{\partial y} + w\frac{\partial u}{\partial z} = -\frac{1}{\rho}\frac{\partial p}{\partial x} + K_h\left(\frac{\partial^2 u}{\partial x^2} + \frac{\partial^2 u}{\partial y^2}\right) + \frac{\partial}{\partial z}\left(K_v\frac{\partial u}{\partial z}\right) + fv$$

$$\frac{\partial v}{\partial t} + u\frac{\partial v}{\partial x} + v\frac{\partial v}{\partial y} + w\frac{\partial v}{\partial z} = -\frac{1}{\rho}\frac{\partial p}{\partial y} + K_h\left(\frac{\partial^2 v}{\partial x^2} + \frac{\partial^2 v}{\partial y^2}\right) + \frac{\partial}{\partial z}\left(K_v\frac{\partial v}{\partial z}\right) - fu.$$

Equation (3.27) describes the hydrostatic approximation, which replaces the equation of motion for vertical currents in most tide–surge models. Some non-hydrostatic models, such as the TRIM3D model (Casulli and Cattani, 1994; Casulli and Stelling, 1998; Kapitza, 2002), may also account for the explicit treatment of vertical movements. This feature may be essential for modeling on very small scales when a steep bathymetry is present. From the hydrostatic relation the pressure at level z is found by integrating (3.27) from the surface η to level z

$$p(z) = p(\eta) + g\int_z^\eta \rho \, dz, \tag{3.29}$$

where $p(\eta)$ corresponds to atmospheric pressure at the free surface. If density is expressed as $\rho = \rho_0 + \rho'$ it follows that $p(z) = p(\eta) + g\rho_0(\eta - z) + g\int_z^\eta \rho' \, dz$. Using this relation, the horizontal pressure gradients in the equation of motion (3.26) become

$$\frac{1}{\rho}\frac{\partial p}{\partial x} = g\frac{\partial \eta}{\partial x} + \frac{g}{\rho_0}\int_z^\eta \frac{\partial \rho'}{\partial x}dz \qquad \text{and} \qquad \frac{1}{\rho}\frac{\partial p}{\partial y} = g\frac{\partial \eta}{\partial y} + \frac{g}{\rho_0}\int_z^\eta \frac{\partial \rho'}{\partial y}dz. \tag{3.30}$$

The first term on the right-hand side describes the barotropic pressure gradient, with $h + \eta$ representing the depth of the water body. The second term is the baroclinic pressure gradient, which has been simplified by assuming that $1/\rho$ may be approximated by $1/\rho_0$. The baroclinic terms depend on the horizontal gradients in water density; that is, on the spatial distribution of temperature and salinity. Contrary to regional and global ocean models, density in tide–surge models is usually not treated in a prognostic manner, but is assumed to be either uniform in space or constant in time so that it can be prescribed. The approach usually works well in shallow and well-mixed continental shelf seas. In the following, baroclinic terms are disregarded.

The third equation (3.28) describes the conservation of mass in an incompressible fluid. Integrating the equation vertically from the bottom $z = -h$ to the free surface η yields a prognostic equation for the free surface:

$$\frac{\partial \eta}{\partial t} + \frac{\partial}{\partial x}\left[\int_{-h}^\eta u\,dz\right] + \frac{\partial}{\partial y}\left[\int_{-h}^\eta v\,dz\right] = \frac{\partial \eta}{\partial t} + \mathbf{V}_h \cdot [(h+\eta)\mathbf{u}_h] = 0 \tag{3.31}$$

with the vertically averaged (or barotropic) horizontal velocity

$$\mathbf{u}_h = \frac{1}{h+\eta}\int_{-h}^\eta \mathbf{u}\,dz. \tag{3.32}$$

For integration of the set of equations, boundary conditions at the free surface and at the bottom of the water body are required. The upper boundary condition is wind stress formulated as

$$K_v \frac{\partial u}{\partial z} = \tau_x \qquad \text{and} \qquad K_v \frac{\partial v}{\partial z} = \tau_y. \tag{3.33}$$

Wind stress $\tau = [\tau_x, \tau_y]$ is parameterized as a function of wind at 10 m height $\mathbf{U}_{10}$ and surface current $\mathbf{u}_s$

$$\tau = c_d \frac{\rho_a}{\rho} (\mathbf{U}_{10} - \mathbf{u}_s) \cdot |\mathbf{U}_{10}|, \tag{3.34}$$

with the drag coefficient either being a constant in the order $c_d = 1.4 \times 10^{-3}$ or depending on wind speed itself. The ratio of atmospheric density ρ_a to water density ρ is often set to a constant value of 1.25×10^{-3}. Usually $\mathbf{U}_{10} \gg \mathbf{u}_s$ such that the current $\mathbf{u}_s$ is almost always disregarded. At the lower boundary bottom friction is implemented by assuming

$$K_v \frac{\partial u}{\partial z} = \gamma u \qquad \text{and} \qquad K_v \frac{\partial v}{\partial z} = \gamma v \text{ with } \gamma = \frac{g \sqrt{u^2 + v^2}}{C_z^2} \tag{3.35}$$

where C_z is the bottom friction coefficient. When the model has open boundaries at its sides, lateral boundary conditions (water levels) need to be supplied as well. These may comprise tides and/or surges that may be obtained from observations, tide models, or ocean models run for larger spatial domains.

3.5.1 Shallow-water equations

When simulating tides, wind-driven sea level elevations, and vertically averaged currents in shallow, well-mixed water bodies, the set of equations just introduced may be further simplified to so-called *shallow-water equations*. In particular, continental shelf or marginal seas with significant tidal amplitudes and tidal mixing satisfy these conditions. An example of such an ocean area is the North Sea. Here Kauker and Langenberg (2000) simulated the sea level response to variable wind forcing in multi-decadal[19] runs with two models, a shallow water model and a three-dimensional regional ocean model (Kauker, 1999) in which baroclinic effects were included. They found that the results from both models were in good agreement and that water level variations were only insignificantly affected by the inclusion of baroclinic factors. In case of other regional seas such as the Baltic Sea, the approach is, however, inadequate because of pronounced vertical stratification and only insignificant tidal variations.

The shallow-water equations have two prognostic variables; namely, vertically averaged horizontal velocity $\mathbf{u}_h$ and surface displacement (water level) η. The

---

[19] Here decadal refers to decades of years.

106 **Models for the marine environment**   [Ch. 3

equation for $\mathbf{u}_h$ takes the form

$$\frac{\partial \mathbf{u}_h}{\partial t} + \mathbf{u} \cdot \mathbf{V}_h \mathbf{u}_h + f\mathbf{k} \times \mathbf{u}_h - K_h \nabla_h^2 \mathbf{u}_h = -g \mathbf{V}_h \eta - \int_{-h}^{\eta} (\tau - R\mathbf{u}_h) \, dz \qquad (3.36)$$

whereas the equation for η is just (3.31). Boundary conditions at the bottom and at the surface are incorporated by the last two terms on the right-hand side of (3.36). The bottom friction coefficient R is usually either given by a nonlinear form of the type $R = c_d \dfrac{\eta}{(h + \eta)} |\mathbf{u}_h|$ or is assumed to be constant. Again the system is closed by lateral boundary conditions

$$\mathbf{u}_h \cdot \mathbf{n} = 0 \qquad \text{and} \qquad \eta = \eta_0(t) \qquad (3.37)$$

at coastlines and at open boundaries, respectively. Here $\mathbf{n}$ represents the unit vector perpendicular to the coastline and $\eta_0(t)$ are prescribed water levels that introduce the influence of the tides and/or surges from adjacent sea areas. Again, the latter may be derived from observations or from model simulations covering larger areas.

3.5.2 Performance of tide–surge models

Tide–surge models are nowadays routinely used for various applications. These range from operational forecasts and case studies to multi-decadal hindcasts covering many decades. Examples of tide–surge models comprise the TRIM3D model (Casulli and Cattani, 1994; Casulli and Stelling, 1998; Kapitza, 2002), TELEMAC2D (Hervouet an Haren, 1996), HAMSOM (Backhaus, 1985) or the POL tide–surge model (e.g., Flather *et al.*, 1998). Within the context of this book mainly applications in which such models have been run to simulate many decades of years are relevant. There have been numerous such applications demonstrating the skill of the models in reconstructing observed long-term tide–surge variations (e.g., Flather *et al.*, 1998; Langenberg *et al.*, 1999; Weisse and Pluess, 2006). One of the more recent integrations is described in Weisse and Pluess (2006). Their calculations were made using the TELEMAC2D model (Hervouet and Haren, 1996) with a computational mesh covering the North Sea consisting of about 50,000 triangular elements and 27,000 nodes. The distances between the nodes vary between about 75 m near the coast and in the estuaries and 27 km in open sea regions. Using high-resolution wind fields that were available every hour, the model simulated the period 1958–2002. Figure 3.9 (see color section) gives an impression of the performance of such a simulation. Shown are comparisons of more than 150,000 simulated and observed hourly values for two tide gauge locations in the German Bight. For both total water levels and surge levels, a high degree of resemblance can be inferred. Correlations are generally well above 0.9 and root-mean-square errors are in the order of about 30 cm for total water levels and about 15 cm for surge levels. Similar exercises have been made by other authors and for other areas. We will return to these in more detail in Section 5.5.

3.6 SUMMARY

Models are an essential tool for the investigation of past and potential future changes of the marine climate. In general, one can distinguish between *cognitive models* that are highly simplified and which are used to conceptualize phenomena, and *quasi-realistic models*. The latter constitute a reality substitute and are used to conduct otherwise impossible experiments. Such experiments comprise the reconstruction of past climate conditions or scenarios for potential future climate changes. Within the context of this book, only quasi-realistic models are considered.

Quasi-realistic models describe only part of the reality. There are always processes unaccounted for in such models and care has to be taken that such models are applied appropriately. Quasi-realistic models cannot be *verified* (i.e., we cannot say whether a model is in general right or wrong), but we can *validate* such models (i.e., we can add to the credibility of the models by analyzing the dynamical system and assuring that all relevant processes are adequately taken into account). However, even comprehensive validation does not provide certainty about the skill of the model when used in different applications.

Quasi-realistic models relevant within the context of this book comprise climate models, wind wave models, and tide–surge models. Here climate model refers to all global and regional as well as ocean and atmosphere models. Such models are not only used to provide scenarios for potential future anthropogenic climate change, but may be applied to a number of other problems as well. These comprise short-term forecasting, the reconstruction of past environmental conditions, dynamical test of hypotheses, and data analysis. When reconstructing past environmental conditions, models are used to provide consistent guesses about past states of the climate system; in particular, about variables for which no measurements have been taken or for areas in which only limited observational evidence is available. The latter is especially important in the context of marine climate and climate change, as the oceans notoriously represent data-sparse regions. This application is closely related to data analysis.

Global climate models may be run either in coupled mode, where an atmospheric and an oceanic component are interactively coupled via the relevant fluxes at the air–sea interface, or in uncoupled mode, where an ocean or an atmosphere model are run standalone. Integration of such models usually is expensive in terms of required computing power. Their spatial resolution thus remains limited. For regional studies, regional models are often used to overcome these constraints and to refine space and time resolution, at least for limited areas. This approach assumes that regional climate is to some extent controlled by climate at larger scales (see Section 1.4.2). Regional climate models are usually run in uncoupled mode driven by boundary conditions derived from global or larger scale regional simulations. So far, only a few exceptions exist, in which regional climate models were run in coupled mode.

Wind wave and tide–surge models are usually applied decoupled from climate model simulations. Only a few exceptions exist in which wind wave models and atmospheric models (e.g., Janssen, 1991; Weber *et al.*. 1993; Weisse and Schneggenburger, 2002) or wind wave and tide–surge models (Choi *et al.*, 2003) have been run in

108 **Models for the marine environment**   [Ch. 3

coupled mode. Usually, however, these models are mostly driven by near-surface wind and pressure fields obtained from climate model simulations and their feedback on the larger scale climate is considered to be small and neglected.

3.7 REFERENCES

Arakawa, A.; and V. Lamb (1977). Computational design of the basic processes of the UCLA general circulation model. *Methods Comput. Phys.*, **17**, 173–265.

Atkinson, K. (1989). *An Introduction to Numerical Analysis*. John Wiley & Sons. ISBN 9780471624899.

Backhaus, J. (1985). A three dimensional model for the simulation of shelf sea dynamics. *D. Hydrogr. Z.*, **38**, 165–187.

Booij, N.; R. Ris; and L. Holthuijsen (1999). A third generation wave model for coastal regions, 1: Model desription and validation. *J. Geophys. Res.*, **104**, 7649–7666.

Bray, D.; and H. von Storch (2007). *The Perspective of Climate Scientists on Global Climate Change*, GKSS Report 2007/11. GKSS Forschungzentrum, Geesthacht, Germany.

Castro, C. R. P., Sr.; and G. Leoncini (2005). Dynamical downscaling: Assessment of value retained and added using the regional atmospheric modeling system (RAMS). *J. Geophys. Ees.*, **110**, D05108, doi: 10.1029/2004JD004721.

Casulli, V.; and E. Cattani (1994). Stability, accuracy and efficiency of a semi-implicit method for three-dimensional shallow water flow. *Computers Math. Applic.*, **27**, 99–112.

Casulli, V.; and G. Stelling (1998). Numerical simulation of 3d quasi-hydrostatic, free-surface flows. *J. Hydr. Eng.*, **124**, 678–698.

Charnock, H. (1955). Wind stress on a water surface. *Quart. J. Roy. Meteorol. Soc.*, **81**, 639–640.

Choi, B.; H. Eum; and S. Woo (2003). A synchronously coupled tide-wave-surge model of the Yellow Sea. *Coastal Eng.*, **47**, 381–398.

Crowley, T.; and G. North (1991). *Paleooclimatology*. Oxford University Press, New York, 330 pp.

Davies, H. (1976). A lateral boundary formulation for multi-level prediction models. *Quart. J. Roy. Meteorol. Soc.*, **102**, 405–418.

Denis, B.; D. C. R. Laprise; and J. Cote (2003). Downscaling ability of one-way nested regional climate models: The big brother experiment. *Climate Dyn.*, **18**, 627–646.

Doyle, J. (1995). Coupled ocean wave–atmosphere mesoscale model simulations of cyclogenesis. *Tellus*, **47A**, 766–778.

Feser, F. (2006). Enhanced detectability of added value in limited-area model results separated into different spatial scales. *Mon. Wea. Rev.*, **134**, 2180–2190.

Flather, R.; J. Smith; J. Richards; C. Bell; and D. Blackman (1998). Direct estimates of extreme storm surge elevations from a 40-year numerical model simulation and from observations. *Global Atmos. Oc. System*, **6**, 165–176.

Giorgi, F.; and L. Mearns (1991). Approaches to the simulation of regional climate change: A review. *Rev. Geophys.*, **29**, 191–216.

Hasselmann, K. (1976). Stochastic climate models, I: Theory. *Tellus*, **28**, 473–485.

Hasselmann, S.; and K. Hasselmann (1985). Computations and parameterizations of the nonlinear energy transfer in a gravity-wave spectrum, Part 1: A new method for efficient computations of the exact nonlinear transfer integral. *J. Phys. Oceanogr.*, **15**, 1369–1377.

Hervouet, J.; and L. V. Haren (1996). *TELEMAC2D Version 3.0 Principle Note*, Rapport EDF HE-4394052B. Electricité de France, Départment Laboratoire National d'Hydraulique, 98 pp.

Hesse, M. (1970). *Models and Analogies in Science*. University of Notre Dame Press, Notre Dame, 184 pp.

Houghton, J.; L. M. Filho; B. Callander; N. Harris; A. Kattenberg; and K. Maskell (1996). *Climate Change 1995: The Science of Climate Change*. Cambridge University Press, 572 pp.

Houghton, J.; Y. Ding; D. Griggs; M. Noguer; P. van der Linden; X. Dai; K. Maskell; and C. Johnson (Eds.) (2001). *Climate Change 2001: The Scientific Basis. Contribution of Working Group I to the Third Assessment Report of the Intergovernmental Panel on Climate Change*. Cambridge University Press, Cambridge, U.K., 881 pp. ISBN 0521 01495 6.

Janssen, P. (1991). Quasi-linear theory of wind wave generation applied to wave forecasting. *J. Phys. Oceanogr.*, **21**, 1631–1642.

Jones, P. (1995). The instrumental data record: Its accuracy and use in attempts to identify the "CO_2 signal". In: H. von Storch and A. Navarra (Eds.), *Analysis of Climate Variability: Applications of Statistical Techniques*. Springer-Verlag.

Kapitza, H. (2002). *Trim Documentation Manual*, Technical Report. GKSS Research Center, Geesthacht, Germany, 51 pp.

Karl, T.; R. Quayle; and P. Groisman (1993). Detecting climate variations and change: New challenges for observing and data management systems. *J. Climate*, **6**, 1481–1494.

Kauker, F. (1999). Regionalization of climate model results for the North Sea. Ph.D. thesis, University of Hamburg, Hamburg, Germany. Available as GKKS Rep. 99/E/6 from GKSS Forschungszentrum, Geesthacht, Germany.

Kauker, K.; and H. Langenberg (2000). Two models for the climate change related development of sea levels in the North Sea: A comparison. *Climate Res.*, **15**, 61–67.

Kharin, V. V.; M. Loussier; X. Zhang; and F.W. Zwiers (2005). Intercomparison of near surface temperature and precipitation extremes in AMIP-2 simulations, reanalyses, and observations. *J. Climate*, **18**, 5201–5223.

Komen, G.; L. Cavaleri; M. Donelan; K. Hasselmann; and P. Janssen (1994). *Dynamics and Modelling of Ocean Waves*. Cambridge University Press, Cambridge, U.K., 532 pp.

Langenberg, H.; A. Pfizenmayer; H. von Storch; and J. Sündermann (1999). Storm-related sea level variations along the North Sea coast: Natural variability and anthropogenic change. *Continental Shelf Res.*, **19**, 821–842.

Lorenz, E. N. (1963). Deterministic nonperiodic flow. *J. Atmos. Sci.*, **20**, 130–141.

Makin, V.; V. Kudryavtsev; and C. Mastenbroek (1995). Drag of the sea surface. *Boundary-Layer Meteorol.*, **73**, 159–182.

McGuffie, K.; and A. Henderson-Sellers (1997). *A Climate Modelling Primer*. John Wiley & Sons, Second Edition, 253 pp.

Miguez-Macho, G.; G. Stenchikov; and A. Robock (2004). Spectral nudging to eliminate the effects of domain position and geometry in regional climate model simulations. *J. Geophys. Res.*, **109**, D13104, doi: 10.1029/2003JD004495.

Miles, J. (1957). On the generation of surface waves by shear flows. *J. Fluid Mech.*, **3**, 185–204.

Müller, P.; and H. von Storch (2004). *Computer Modelling in Atmospheric and Oceanic Sciences: Building Knowledge*. Springer-Verlag, 304 pp.

Navarra, A. (1995). The development of climate research. In: H. von Storch and A. Navarra (Eds.), *Analysis of Climate Variability: Applications of Statistical Techniques*, GKSS School on Environmental Science. Springer-Verlag.

Nieto Borge, J.; G. Rodríguez; K. Hessner; and P. González (2004). Inversion of marine radar images for surface wave analysis. *J. Atmos. Oceanic Technol.*, **21**, 1291–1300.

Okasha, S. (2002). *Philosophy of Science*. Oxford University Press. ISBN 9780192802835.

Oreskes, N.; K. Shrader-Frechette; and K. Beltz (1994). Verification, validation, and confirmation of numerical models in earth sciences. *Science*, **263**, 641–646.

Phillips, O. (1957). On the generation of waves by a turbulent wind. *J. Fluid Mech.*, **2**, 417–445.

Pielke, Sr., R. (1991). A recommended specific definition of "resolution". *Bull. Am. Meteorol. Soc.*, **72**, 1914.

Räisänen, J.; U. Hansson; A. Ullerstig; R. Döscher; L. Graham; C. Jones; H. Meier; P. Samuelsson; and U. Willén (2004). European climate in the late twenty-first century: Regional simulations with two driving global models and two forcing scenarios. *Climate Dyn.*, **22**, 13–31, doi: 10.1007/s00382-003-0365-x.

Rinke, A.; and K. Dethloff (2000). On the sensitivity of a regional Arctic climate model to initial and boundary conditions. *Climate Res.*, **14**, 101–113.

Robinson, A. R.; P. F. J. Lermusiaux; and N. Q. Sloan (1998). Data assimilation: In: K. Brink and A. Robinson (Eds.), *The Global Coastal Ocean: Processes and Methods*, The Sea Vol. 10. John Wiley & Sons, New York.

Schneggenburger, C.; H. Günther; and W. Rosenthal (1997). Shallow water wave modelling with nonlinear dissipation. *D. Hydrogr. Z.*, **49**, 431–444.

Solomon, S.; D. Qin; M. Manning; Z. Chen; M. Marquis; K. Averyt; M. Tignor; and H. Miller (Eds.) (2007). *Climate Change 2007: The Physical Science Basis. Contribution of Working Group I to the Fourth Assessment Report of the Intergovernmental Panel on Climate Change*. Cambridge University Press, U.K., 996 pp. ISBN 978-0-521-88009-1.

Suarez, M.; and P. Schopf (1988). A delayed action oscillator for ENSO. *J. Atmos. Sci.*, **45**.

Tolman, H. (1991). A third-generation model for wind waves on slowly varying, unsteady and inhomogeneous depths and currents. *J. Phys. Oceanogr.*, **21**, 782–797.

Trenberth, K. (1993). *Climate System Modeling*. Cambridge University Press, 788 pp.

Trigo, R.; and J. Palutikof (1999). Simulation of daily temperatures for climate change scenarios over Portugal: A neural network model approach. *Climate Res.*, **13**, 1–75.

von Storch, H. (1995). Inconsistencies at the interface of climate impact studies and global climate research. *Meteor. Z.*, **4**, 72–80.

von Storch, H. (2001). Models between academia and applications. In: H. von Storch and G. Flöser (Eds.), *Models in Environmental Research*, GKSS School on Environmental Science. Springer-Verlag, pp. 17–33.

von Storch, H.; and G. Flöser (2001). *Models in Environmental Research*, GKSS School on Environmental Science. Springer-Verlag, 254 pp.

von Storch, H.; S. Güss; and M. Heimann (1999). *Das Klimasystem und seine Modellierung: Eine Einführung*. Springer-Verlag [in German]. ISBN 978-3-540-65830-6.

von Storch, H.; H. Langenberg; and F. Feser (2000). A spectral nudging technique for dynamical downscaling purposes. *Mon. Wea. Rev.*, **128**, 3664-3673.

von Storch, H.; J.-S. von Storch; and P. Müller (2001). Noise in the climate system: Ubiquitous, constitutive and concealing. In: B. Engquist and W. Schmid (Eds.), *Mathematics Unlimited: 2001 and Beyond*, Part II, GKSS School on Environmental Science. Springer-Verlag.

Waldron, K. M.; J. Peagle; and J. Horel (1996). Sensitivity of a spectrally filtered and nudged limited area model to outer model options. *Mon. Wea. Rev.*, **124**, 529–547.

WAMDI-Group (1988). The WAM model: A third generation ocean wave prediction model. *J. Phys. Oceanogr.*, **18**, 1776–1810.

Washington, W.; and C. Parkinson (1986). *An Introduction to Three-dimensional Climate Modelling*. University Science Books, 422 pp.

Weber, S.; H. von Storch; P. Viterbo; and L. Zambresky (1993). Coupling an ocean wave model to an atmospheric general circulation model. *Climate Dyn.*, **9**, 63–69.

Weisse, R.; and F. Feser (2003). Evaluation of a method to reduce uncertainty in wind hindcasts performed with regional atmosphere models. *Coastal Eng.*, **48**, 211–225.

Weisse, R.; and A. Pluess (2006). Storm-related sea level variations along the North Sea coast as simulated by a high-resolution model 1958–2002. *Ocean Dynamics*, **56**, 16–25, doi: 10.1007/s10236-005-0037-y, online 2005.

Weisse, R.; and C. Schneggenburger (2002). The effect of different sea state dependent roughness parameterizations on the sensitivity of the atmospheric circulation in a regional model. *Mon. Wea. Rev.*, **130**, 1595–1602.

Weisse, R.; H. Heyen; and H. von Storch (2000). Sensitivity of a regional atmospheric model to a sea state dependent roughness and the need of ensemble calculations. *Mon. Wea. Rev.*, **128**, 3631–3642.

WMO (1998). *Guide to Wave Analysis and Forecasting*, Vol. WMO 702. World Meteorological Organization. ISBN 92-63-12702-6.

Zahel, W. (1978). The influence of solid earth deformations on semidiurnal and diurnal oceanic tides. In: P. Brosche and J. Sündermann (Eds.), *Tidal Friction and the Earth's Rotation*. Springer-Verlag.

4

How to determine long-term changes in marine climate

4.1 INTRODUCTION

So far we have reviewed the dynamics of the global climate system, the marine weather phenomena this book is about—in particular, storms, wind waves, and storm surges—and how to mathematically describe these phenomena. In this chapter, we address the question on how to determine *long-term changes* in the *statistics* of marine weather phenomena.

The expression *long-term* is only loosely defined. In different scientific areas it refers to different time scales. For example, geologists may refer to long-term changes when variations on time scales of millennia are considered. In atmospheric research, the meaning varies from hours to thousands of years when atmospheric turbulence or paleoclimatic changes are looked at. Within this book, we use the phrase *long-term* to refer to changes that occur over decades; that is, the time horizon accessible to human experience. In some cases, a century or two may be considered.

Time scales of decades are of outstanding relevance for many marine practitioners with responsibilities for coastal and offshore management. For them, it is not enough just to be informed about conditions right now and about changes to be expected within the next few hours. The relevant question is: What are the long-term changes and how long will these changes continue into the future? For example, if we observe a season with numerous and intense hurricanes, such as 2005, the conclusions to be drawn and the practical implications are different when we perceive this season as just one step in an ongoing series of intensification extending into the future, or merely as one extraordinary season, with the following seasons being more comparable with those in earlier years. In the latter case, the conclusion to be drawn for a coastal manager is mainly to keep the present strategy, but to review and possibly adjust the present levels of dams while at the same time keeping them in order. In the former case, we may have to think about completely different strategies,

114 **How to determine long-term changes in marine climate?** [Ch. 4

such as whether occupation and rebuilding in low-lying areas should be managed more restrictively.

Assessing whether a change is of intermittent or of ongoing character requires the identification of causes. Is the change within the range of natural variations which we see from season to season or from decade to decade, or is it related to a *driver* which is expected to operate into the future as well? Examples, beyond the obvious case of anthropogenic climate change, are land sinking in coastal areas due to oil and gas exploration, and elevated storm surge levels in estuaries related to the deepening of shipping channels.

In the following we will discuss some concepts that may be used to address these sorts of questions and the problems and caveats associated with them. We begin with a discussion of problems related to data quality and focus on issues that have a high potential for making wrong inferences about long-term changes (Section 4.2). In Section 4.3 we introduce the concept of proxy data that may be used when the phenomenon or variable of interest has not really been observed, or when observational evidence is compromised by short instrumental records or data inhomogeneities (see Section 4.2.1). Reanalyses and reconstructions are introduced in Section 4.4. Here advanced numerical models are used to provide an optimal space–time interpolation of the observational record to derive a dynamically consistent picture of the observed state, as well as information on variables for which no measurements have been made. We discuss the extent to which long-term changes (trends) can be estimated from such data and review major reanalysis and reconstruction projects. Regionalization techniques are described in Section 4.5. Their main purpose is to derive regional details from large or global-scale information; for example, what can we infer about long-term changes in the sea state statistics for an oil field in the North Sea, if only variations of the large-scale atmospheric pressure field are known. The concept of scenarios and projections is introduced in Section 4.6. The concept is widely used in marine and climate sciences to address questions of the sort "What ... if ...?", to elaborate on their consequences, and to develop strategies to deal with them. We conclude with a brief introduction of detection and attribution (Section 4.7). Detection refers to techniques used to decide whether or not an observed change is unusual in some sense, while attribution refers to assigning causes to these changes. In particular, attribution to causes is required when statements about a future continuation of a change are needed. If developments as a result of these causes are predictable, such as the isostatic rebound of the land in some coastal regions, then we are also able to predict, at least to some extent, the consequences for our phenomenon of interest.

4.2 PROBLEMS WITH DATA QUALITY

In order to determine and to assess long-term changes in marine climate, knowledge about the state of the system in its entirety is required. It is, however, fundamentally impossible to obtain such information. Instead, we rely on *samples* of data that provide a more or less complete or scattered picture about the system's state. In a statistical sense, it is assumed that *random samples* are drawn from a *population* (the system's state) in order to make *inferences*.

It is clear that the concept of sampling geophysical processes is complex and that strong assumptions are implicit in applying the conceptual model of *random sampling* to the analysis of climate data. For example, the concept requires that repeatable experiments are performed in which elements from the sample space are drawn randomly and with replacement (von Storch and Zwiers, 1999). Suppose that we want to determine the mean temperature of the North Sea from a limited number of observations. Naturally, measurements will be restricted to a certain time interval, to some measurement points, and to a certain frequency at which the measurements were repeated (*sampling interval*). There will be no data from the distant past or the future, from North Sea locations at which no measurements have been taken, or in-between the interval at which the observations were made. In order to treat the sample as a random sample and to make inferences about the system's state some assumptions about underlying processes are therefore made implicitly: it is assumed that the processes are *stationary* or *cyclo-stationary* (i.e., their statistical properties are time-invariant), and that the processes are *ergodic* (i.e., sampling a given realization of a process in time yields information equivalent to randomly sampling independent realizations of the same process; von Storch and Zwiers, 1999). Obviously, these are sometimes strong assumptions for sampling geophysical processes and careful sampling strategies are required to make appropriate inferences. In the following, we review some problems related to sampling strategies and their consequences for the interpretation of long-term changes in the marine climate. In Section 4.2.1 we first introduce the problem of data homogeneity in general. Subsequently in Sections 4.2.2 and 4.2.3 some specific homogeneity problems related to changing data availability and the interpretation of data from derived products are reviewed.

4.2.1 Data homogeneity

When measurements are taken the outcome depends on a number of factors. The *true value* of the measured parameter at the specific time and place is of course important, but factors such as the type or the accuracy of the instrument also play an important role. When measurement errors or practices have changed in the course of time and are not adequately accounted for, they may have a profound influence on the interpretation of long-term changes and may lead to wrong inferences. For example, in earlier times marine wind speeds were derived from visual assessments of the sea state and were reported according to the Beaufort Scale. The accuracy and also the outcome of these assessments depend on a number of factors such as the experience of the sailor who made the observation, or the type and the speed of the vessel relative to the observed sea state. Also, the accuracy of this type of data is clearly different from modern wind speed data derived from satellite retrievals of radar waves reflected from the sea surface (see Figure 2.11). Both types of observations are very useful for different purposes, but they are not *comparable per se*. The latter may represent a problem when both types of data are mixed to construct long time series to make inferences about long-term changes in marine wind speeds. In this case, it is not clear whether winds today are indeed stronger or weaker or if we merely look at differences due to the different sampling strategies. While the problem appears to be rather obvious, it may be difficult to track when both types of data are combined into a derived product (see Section 4.2.3) and when only limited information about the original data remains available. For instance, both types of data may be combined into a monthly mean product with a smaller/higher number of visual observations contributing in earlier/more recent years. In this case, a long-term trend in the derived product may be generated which is not a manifestation of long-term climate change, but merely a manifestation of changing observational practices and sampling strategies.

Even if the measurements are correct and the technique has not changed in the course of time similar problems may arise. When daily mean temperatures are considered, it matters significantly if they were derived from averaging three observations taken at 06:00, 12:00 and 18:00 UTC or at 07:00, 13:00 and 19:00 UTC. Here the derived daily means could not be compared because they describe different quantities. There are a number of other factors that may cause similar problems. Changes in the exposure of the instrument may play a role. Examples reported in the literature describe vertical displacements of rain and snow gauges which resulted in massive changes of recorded rainfall and snowfall (e.g., Karl et al., 1993), or changes in the surroundings of weather stations that lead to changes in recorded wind speed. Another example is provided by a controversy that raged in the 1990s, where in situ temperature measurements taken by radiosondes throughout the troposphere indicated a warming trend, while at the same time satellite-based temperatures showed no such trend. The puzzle was later solved in favor of the in situ data when it became clear that changes in the satellite tracks in the course of time had influenced satellite retrievals, so that the satellite-based temperature data reflected not only information about the temperature but also aspects of the flight track.

In principle long-term changes inferred from historical observational records may thus originate from two different sources: They may reflect either changes in the statistics of the parameter actually monitored (i.e., the *signal*), or problems related to changing instrumental accuracy, observational practices, analysis routines, etc. In the following we refer to the latter as *inhomogeneities* and call a data set *homogeneous* if it is free from such contaminations.

Karl et al. (1993) characterized inhomogeneities as being either *creeping* or *sudden*. Sudden inhomogeneities are introduced by abrupt, but often documented changes in the sampling strategy. Such changes comprise the replacement or improvement of instruments, the upgrading from manual to automatic analysis techniques (e.g., Günther et al., 1998), or changes in the density of the observational network in relatively data-sparse regions such as oceans. When sudden inhomogeneities are not already known from documentation, they may often be identified from screening the time series for jumps in the moments calculated for moving time windows (WASA-Group, 1998). In Figure 4.1 two examples of sudden inhomogeneities are illustrated. The first example shows the number of days per year with wind speeds of 7 Beaufort or more in the city of Hamburg, northern Germany. Obviously, a noticeable decrease in the number of these days occurred around about

Sec. 4.2]	Problems with data quality 117

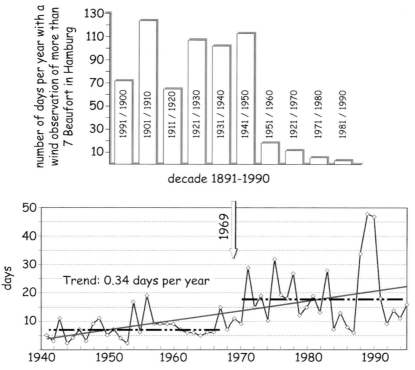

Figure 4.1. Number of days per decade where wind speed observations at noon exceeded 7 Beaufort or more in Hamburg, Germany (top); courtesy of Heiner Schmidt, German National Weather Service. Number of days per year where wind speeds of 21 m s^{-1} or more (storm days) have been observed in Kullaberg, southwest Sweden (bottom); The solid line represents annual values, the straight solid line the linear trend over the time interval shown, and the two horizontal lines are the 1941–1969 and the 1970–1995 averages, respectively.

1950, the explanation for which is relocation of the instrument from the harbor to the city airport. The second example is less obvious and has occasionally been mistaken as evidence for an intensification of storm activity in northern Europe. It shows the number of storm days per year in Kullaberg, southwest Sweden, from 1940 onwards. At a first glance, the curve exhibits a linear trend with storm events being more frequent in the later part of the time series. However, indications do exist that in 1969 a major storm severely damaged the forest surrounding the measurement site, so that locally recorded wind speeds may have become stronger after the wind break as a result of decreased surface roughness and increased exposure of the instrument (von Storch and Weisse, 2008). When the two periods before and after 1969 are considered independently, the number of storm days remained relatively constant within each period, although the mean level is somewhat higher in the later period.

Creeping inhomogeneities are usually more difficult to detect and bear a particularly high potential for wrong inferences. They may originate from ongoing

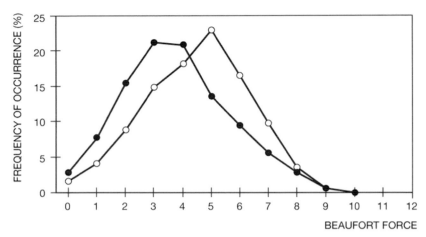

Figure 4.2. Frequency distribution of wind estimates reported by ships of opportunity; distribution estimated from 24,442 visual reports (solid circles) and distribution derived from about 1,000 visual reports in which the observer was supported by an instrument aboard the ship (open circles). Redrawn after Peterson and Hasse (1987).

modifications of the observational network such as changes in the density of the station network or, for marine weather statistics, from gradual changes in ship routing and observational practices on voluntary observing ships. Figure 4.2 illustrates an example. Shown are two wind speed distributions derived from ship records in the English Channel (Peterson and Hasse, 1987). In both cases the reports were made by sailors based on visual assessments according to the Beaufort Scale. One distribution was derived from reports where the observer was assisted by an additional instrument aboard the vessel, the other distribution was derived from cases where no such instrument was available. Obviously, the presence of an additional instrument led the sailors to report higher wind speeds. Since in the course of time more instruments have been mounted aboard the various vessels, such effects may have introduced creeping inhomogeneity into the observational record.

A way of identifying such creeping inhomogeneities is to compare the suspected time series with data from nearby stations that are known to be unaffected (Alexandersson and Moberg, 1997; Moberg and Alexandersson, 1997; Auer *et al.*, 2005). However, particularly for the marine environment such data are often not available. Thus, the methodological challenge in all analyses of long-term marine climate change is to discriminate between *signals* and *inhomogeneities*. In the following we will review some specific homogeneity problems that are related to changing data availability and to the interpretation of data from derived products.

4.2.2 Data availability

Changing data availability in the course of time may cause inhomogeneities especially when derived products are considered. An example is provided in Figure 4.3. Here

Sec. 4.2] Problems with data quality 119

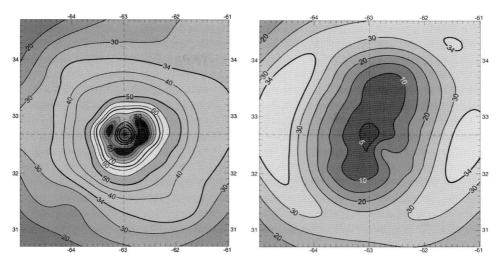

Figure 4.3. Surface wind field analysis of Hurricane Erin at 19:30 UTC on September 9, 2001 with (left) and without (right) the benefit from aircraft reconnaissance data. For details see Landsea et al. (2004). Courtesy of Christopher Landsea, NOAA.

two analyses of Hurricane Erin in 2001 are shown. In one analysis only surface-based measurements were used. In the other, data from aircraft reconnaissance flights were included additionally. Obviously, there is a bias between both analyses with higher maximum sustained wind speeds in the case where aircraft reconnaissance data were used. Aircraft reconnaissance flights emerged in about the mid-1940s. Thus, the first analysis would be typical for the years before 1945, the second for the years after about 1945. As aircraft reconnaissance data have been increasingly incorporated into operational analyses, this may force a trend towards more intense hurricanes in the operational data and may lead to wrong inferences when not adequately accounted for (Landsea et al., 2004).

A similar case for the number of hurricanes is illustrated in Landsea (2007). Here two of the most active hurricane seasons on record (1933 and 2005) before and after the advent of geostationary satellite imagery in 1966 are compared. Landsea (2007) realized that in 2005 a considerably higher number of hurricanes that eventually stayed over the ocean was reported, while the number of landfalling systems was comparable in both years. Splitting the yearly number of tropical storms into the fraction that struck land and into the fraction that remained over the open ocean, he found that more of the latter were observed in recent decades while the number of systems that eventually made landfall remained relatively constant, with some decadal variability superimposed. When Landsea (2007) computed the ratio between the two fractions he found a noticeable discontinuity in the time series that occurred after the advance of geostationary satellite imagery in 1966 (Figure 4.4). He concluded that there may be a substantial undercount in historical hurricane records prior to the advances of satellite observations and that part of the observed upward

120   How to determine long-term changes in marine climate?   [Ch. 4

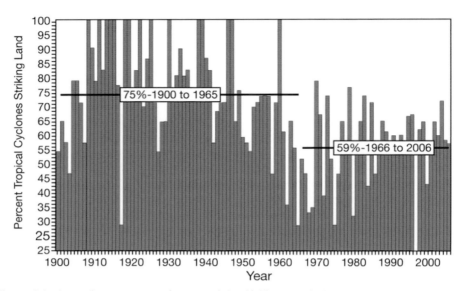

Figure 4.4. Annual percentage of reported landfalling tropical cyclones 1900–2006 in the Atlantic U.S. Redrawn after Landsea (2007). Courtesy of Christopher Landsea, NOAA. Published/copyright (2007) American Geophysical Union. Reproduced/modified by permission of American Geophysical Union.

trend in hurricane frequencies may be attributed to such types of inhomogeneity in the historical records. A similar conclusion was put forward by Shepherd and Knutson (2007).

Figure 4.5 illustrates a similar case for extra-tropical cyclones. In about 1990, available German weather maps for the North Atlantic and Northern Europe were used to count the number of storm systems present in these weather maps and to make inferences about their long-term change. A steady increase in the number of strong wind storms was obtained and exploited as evidence of unfolding anthropogenic climate change. Closer inspection of the data revealed, however, that the data set was inhomogeneous and that the upward trend largely reflected increasing data availability in the course of time. In particular, it was found that the increase mainly occurred in two phases, first in the years immediately after World War II where the density of the observational network in vast oceanic regions was greatly improved, and second almost suddenly around 1970 when satellite imagery became available which allowed almost certain detection of cloud patterns associated with storms. In-between and after these changes, the number of detected cyclones appears almost stationary. Again, the data sets from the different periods are not comparable and doing so may lead to wrong inferences regarding long-term changes in extra-tropical storm activity. It is thus essential that the chosen sampling strategy is adequate; in particular, that the relevant space and time scales are considered and that inhomogeneities that may arise from changing data availability are reduced as far as possible.

Sec. 4.2]					Problems with data quality 121

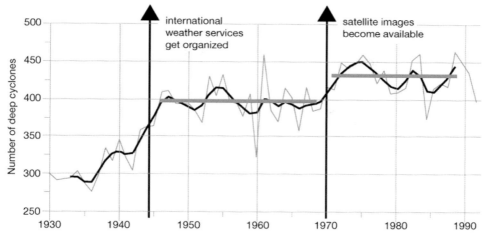

Figure 4.5. Number of deep cyclones detected in daily German weather maps of the North Atlantic.

4.2.3 Operational weather analyses and other derived data sets

The examples in the previous section demonstrated that the homogeneity problem is not limited to observational data, but also often present in derived data sets or products such as weather analyses or reanalyses. In meteorology (and more recently also in oceanography) an *analysis* refers to a procedure to project the state of the atmosphere or the ocean as known from a finite set of imperfect, irregularly distributed observations onto a regular grid or to represent it by the amplitude of standard mathematical functions (Glickman, 2000). The purpose of an analysis is to provide the initial fields for subsequent forecasts by numerical weather prediction models or to use it for easy diagnostic studies. An analysis may be considered as a space–time interpolation system which provides physically *consistent*[1] information about the state of the system and about variables and parameters that have either not been measured directly or not been measured at the same space–time resolution as in measurements that are used as input for the analysis.

Originally, operational analyses were aimed at generating the initial fields for subsequent numerical (weather) forecasts. Because of their gridded nature and their simple accessibility their use for other purposes, such as analysis of long-term changes, quickly became popular. As the preparation of an analysis is carried out by means of sophisticated numerical (weather prediction) models and data

---

[1] Observations are said to be consistent if they are conformable to fundamental physical equations. For example, wind speed is physically linked with the pressure gradient. When measurement errors lead to inconsistencies between pressure and wind measurements the observations are said to be inconsistent. An analysis may to some extent correct for such errors, providing consistent but not necessarily "correct" numbers.

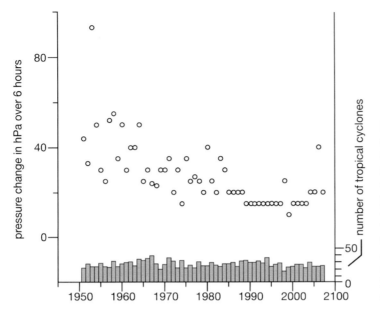

Figure 4.6. Annual number of tropical storms in east Asia (gray) and maximum decrease of central pressure within 6 hours (dots) along a tropical storm track according to the JMA best track data. Note the 93 hPa per 6 hours change in 1953.

assimilation techniques that have matured and changed significantly in recent decades, long-term data obtained from operational analyses are likely to be compromised by inhomogeneities. Similarly, changing data availability may become an issue (see Section 4.2.2). As an example let us consider the monitoring of tropical cyclones and their compilation into so-called *best track* data sets. These data sets contain information about each individual tropical cyclone such as the name of the storm, its intensity or category, the maximum sustained wind speed, the position of the storm's center, the core pressure, etc. They are available from different meteorological services and institutes such as the Japan Meteorological Agency[2] (JMA), the China Meteorological Administration/Shanghai Typhoon Institute,[3] the Joint Typhoon Warning Center,[4] or the U.S. National Hurricane Center.[5] Nowadays best track data date back several decades. In many cases such data sets are obviously very useful, but their homogeneity may be compromised by the changing availability of observational platforms, such as aircraft reconnaissance flights or satellites, or by changing analysis techniques. Figure 4.6 shows the annual number of tropical storms in the northwest Pacific as recorded in the JMA best track data set. The series, which commenced in 1951 and extended until 2007, looks quite stationary with substantial variations from year to year. It is plausible that most if not all of the storms have been detected through the entire period 1951–2007. However, when the maximum pressure

[2] See *http://www.jma.go.jp/jma/jma-eng/jma-center/rsmc-hp-pub-eg/trackarchives.html*
[3] See *http://www.typhoon.gov.cn/en/data/detail.php?id=38*
[4] *https://metocph.nmci.navy.mil/jtwc/best_tracks/*
[5] *http://www.nhc.noaa.gov/pastall.shtml*

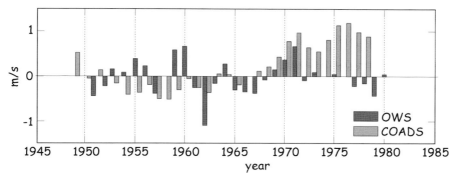

Figure 4.7. Annual mean wind speed anomalies in the North Pacific in the area of ocean weather ship *P* derived from ICOADS data (light gray) and from measurements aboard *P* (dark gray). Courtesy of Hans-Jörg Isemer, GKSS Research Center.

decrease within 6 hours across all tracks within a season is analyzed, substantial changes from very high numbers (as high as 93 hPa within 6 hours in 1953) to suspiciously small and intermittently constant numbers (only 15 hPa within 6 hours during the 1990s) are found. While there still might exist a physical explanation for part of the changes, it nevertheless remains very likely that the record itself is inhomogeneous.

The ICOADS[6] data set represents another type of a derived data product. It consists of a collection of weather reports made by ships, mostly by ships of opportunity, which have been edited and statistically summarized. The ICOADS data comprise measurements of temperature, humidity, wind, surface pressure, sea state, etc. since 1784 and have been widely used in different studies such as in heat balance studies, in ENSO[7] or North Atlantic Oscillation studies, or in studies dealing with satellite calibration, four-dimensional analysis of ocean circulation, shelf sea dynamics, coastal morphology, sea level changes, etc. Figure 4.7 compares annual mean wind speed anomalies from the ICOADS data set with measurements taken aboard the ocean weather ship *P* in the North Pacific. While measurements taken directly aboard the weather ship do not show any significant long-term change apart from some multi-decadal variability, the data taken from voluntary observing ships in the vicinity of the weather ship indicate an upward trend, most pronounced after about 1960. While the ocean weather ship reported quality controlled and mostly homogeneous data throughout the period, some creeping inhomogeneity must be present in the ICOADS data. Such inhomogeneities may result from gradual changes in ship routing, efforts and procedures to avoid severe weather, and observational practices aboard voluntary observing ships. Up to now some work on homogenizing ICOADS data has been performed, but care is still needed when working with this nevertheless excellent data base.

---

[6] The International Comprehensive Ocean–Atmosphere Data Set Project; *http://www.ncdc.noaa.gov/oa/climate/coads/*
[7] El Niño Southern Oscillation.

4.3 PROXY DATA

In previous sections we introduced the concept of *data homogeneity* and illustrated how inhomogeneities may lead to wrong inferences about long-term changes and trends. Homogeneity problems exist for many observational variables. Wind measurements, especially, are notoriously inhomogeneous as they may be affected by a large number of factors that introduce variations which are unrelated to changes and variations in wind climate. Such factors comprise, for instance, changes in the surroundings of the instrument such as growing trees, replacement of the sensor, changes in measuring techniques, or relocation of the measurement site. Some examples are described in more detail in Sections 4.2.1 and 4.2.2.

The obvious question arises as to whether there exist other variables that carry some information about long-term changes in the parameter of interest, but which are less affected by homogeneity problems. Such variables are referred to as *proxy variables* and their data as *proxy data*. Usually, proxy variables are physically linked to the parameter of interest. For instance, pressure gradients may be used as a proxy for wind speed as both are linked via the geostrophic relation (see Appendix A.2). Generally, such information may be derived only for certain locations while no or only limited information about spatial distributions is obtained. Moreover, proxies do not provide information in absolute scales, such as wind speed in $m\,s^{-1}$. Proxies only inform about relative changes; in particular, about increases, decreases, and long-term variations. As such, proxies may only provide information about changing statistics in time.

For storminess, the idea and concepts for proxies were mostly developed in the mid-1990s within the European project WASA.[8] The proxies derived were mainly based on air pressure readings and water levels obtained from tide gauge records. In particular, air pressure readings from synoptic meteorological stations were considered to be relatively homogeneous, as the methodology of measuring surface air pressure has changed very little over the last few centuries and because such measurements are less affected by changes in the surroundings of the measurement site. Tide gauge data, on the other hand, may show a number of local signals not related to storminess, such as long-term sea level rise, changes caused by local bathymetry modifications,[9] or relocation of the tide gauge. Usually, such effects could, however, be accounted for in the design of proxies for storminess (e.g., von Storch and Reichardt, 1997). Both sources of proxy data—air pressure and tide gauge measurements—are available for more than a century at many places. The latter enables us to use these data in the assessment of long-term changes in the underlying storm climate.

An example of a storm activity proxy based on air pressure readings, which is now widely accepted and used in storm climate studies, was suggested by Schmidt and

---

[8] Changing waves and storms in the northeast Atlantic, see *http://coast.gkss.de/staff/storch/projects/wasa.html*

[9] These may be caused either naturally (e.g., by erosion or sedimentation) or anthropogenically (e.g., by dredging).

Sec. 4.3] Proxy data 125

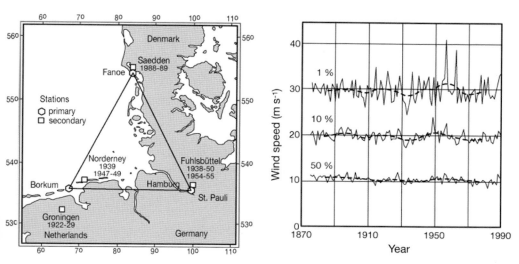

Figure 4.8. Location of meteorological stations from which pressure readings have been used to estimate geostrophic wind speed in the German Bight (left) and time series of the 1-percentile, 10-percentile, and 50-percentile (from top to bottom) geostrophic wind speed derived from annual distributions of daily geostrophic wind speeds in the German Bight (right). Reprinted/adapted by permission from Macmillan Publishers Ltd.: *Nature*, Schmidt and von Storch (1993), copyright 1993.

von Storch (1993). In this article, first daily geostrophic wind speeds[10] were calculated from surface air pressure measurements obtained from a triangle of synoptic stations surrounding the German Bight, North Sea (Figure 4.8, left). In this way a frequency distribution of daily geostrophic wind speeds was obtained for each year, from which subsequently upper percentiles[11] were determined and used as a proxy for storm activity over the German Bight (Figure 4.8, right). Figure 4.9 demonstrates that such a proxy indeed reflects observed wind and storm conditions within a region. It shows a comparison between percentiles derived from geostrophic wind speed estimates and local wind observations at five Danish stations that are known to be quite homogeneous for the 5-year period 1980–1984. A remarkably linear link is found that suggests that a change in the distribution of local upper-wind speed percentiles would be reflected in a corresponding change of the geostrophic wind speed index and *vice versa* (WASA-Group, 1998).

A time series of geostrophic wind speed indices can be usefully employed as a proxy for the detection of changing wind and storm conditions over time. The approach is now well established. Schmidt and von Storch (1993) provided the first study that utilized the approach for the assessment of long-term changes in the marine storm climate over the German Bight. Alexandersson *et al.* (1998) and later Alexandersson *et al.* (2000) used the approach to analyze data from many different

---

[10] For the concept of geostrophic wind speed see Appendix A.2.
[11] Typically, the 95-percentile or 99-percentile is used.

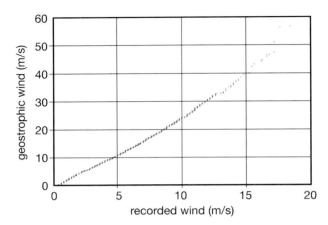

Figure 4.9. Percentile–percentile plot of observed (recorded) daily near-surface wind speed averaged over five synoptic stations in Denmark for which homogeneous wind measurements were available for 1980–1984 and geostrophic wind speeds derived from station pressure data. Redrawn after WASA-Group (1998). Reproduced/modified by permission of American Meteorological Society.

pressure triangles. By averaging over many of these triangles they were able to provide indices that described variations in the storm climate over somewhat larger areas; namely, the North Sea and the Baltic Sea, respectively. An update of their analyses can be found in the *Fourth Assessment Report of the Intergovernmental Panel on Climate Change* (IPCC) (Trenberth et al., 2007). In methodologically similar efforts Matulla and von Storch (2009) and Matulla et al. (2008) provided estimates of long-term changes in eastern Canadian, and northern and central European storminess, respectively. The results of these exercises are given in Chapter 5.

A disadvantage of the pressure triangle–geostrophic wind speed approach is that long and homogeneous pressure data from at least *three* different stations are needed. Further, the distance between the stations should be such that it reasonably reflects the spatial scale of synoptic low-pressure systems; that is, the stations should be neither too remote nor too close to each other. This considerably limits the number of cases for which the approach is applicable and does not fully exploit existing observations. For example, a very small number of stations exists that have very long observational records that date back several hundreds of years. For the very early periods of these records, there are no additional stations available that can be used to complement these observations for the calculation of geostrophic wind speeds. In order to fully exploit the existing data, considerable efforts have therefore been made to develop storm indices that are based on observations from only *one* station. Examples of such indices comprise simple storm count indices such as the annual number of barometer readings below a given threshold (e.g., 980 hPa), the annual number of rapid negative pressure tendencies, or upper percentiles of the latter (e.g., Bärring and von Storch, 2004). While the number of low barometer readings reflects the low pressure usually associated with storm systems, pressure tendency indices focus more on the pressure gradient associated with a propagating cyclone passing the observer. The different indices thus cover different aspects of storm activity and there is no need for them to be strongly linked.

An example of an analysis of storminess proxies based on data from single stations is provided in Bärring and von Storch (2004). They analyzed air pressure

Sec. 4.3] Proxy data 127

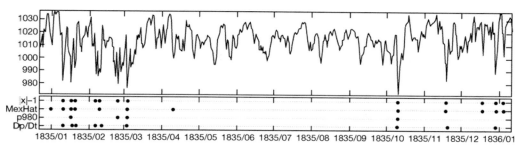

Figure 4.10. Pressure record of 1835 for Lund, Sweden (solid) and detection performance of four different storm indices (dots). The indices are labeled $|x| - 1$, *MexHat*, *p980*, and Dp/Dt. For details see Bärring and Fortuniak (2009). Redrawn after Bärring and Fortuniak (2009).

data from two Swedish stations (Lund and Stockholm) commencing in the late 18th century and early 19th century, respectively. Bärring and von Storch (2004) calculated and compared a number of different indices covering different aspects of storm activity. In general, they found the different proxies to be positively correlated, especially with respect to the absence of any systematic long-term change. However, they did not focus specifically on the differences among the various indices. Using the same data, Bärring and Fortuniak (2009) therefore compared eight different storm indices and showed that the various proxies are indeed far from being identical especially on short time scales. Figure 4.10 shows an example of their analyses for Lund. It illustrates how different proxies may detect different types of cyclones and thus do cover different aspects of storm activity. Nevertheless, on longer time scales all indices are positively correlated, suggesting that there is some large-scale information on cyclone activity that is shared between the different storminess indices. Bärring and Fortuniak (2009) performed an EOF analysis[12] to separate the leading information of inter-annual variability common to all indices. They found that the latter is, at least on decadal time scales, well correlated with the information contained in indices based on the pressure triangle–geostrophic wind speed approach (Figure 4.11). The latter suggests that single-station indices may very well be used to obtain information broadly consistent with what is obtained from multi-station indices, but which may cover longer time periods. A similar comparison within the framework of a global reanalysis (see Section 4.4) was performed by Paciorek et al. (2002). By comparing six different storm indices over different geographical areas they concluded that the different indices provide reasonably consistent measures of cyclone activity in the regions of extra-tropical storm tracks.

Water level variations from tide gauge records may provide an alternative source for the development of storminess proxies. This was first suggested by de Ronde (WASA-Group, 1998). Although local water level variations at tide gauges are often influenced by a number of factors unrelated to storm activity such as local water works, global mean sea level rise, or geological phenomena such as land subsidence or

---

[12] Empirical orthogonal functions, see Zwiers and von Storch (2004) for details.

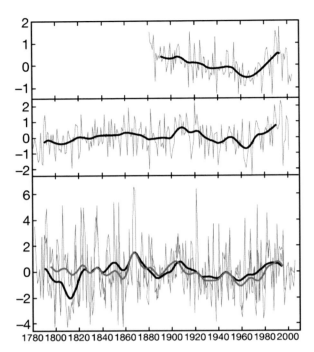

Figure 4.11. Storm index for northern Europe after Alexandersson et al. (1998) (top), NAO index after Luterbacher et al. (2002) (middle), first principle component of multiple storm indices for Lund and Stockholm (bottom). Redrawn after Bärring and Fortuniak (2009).

uplift, these factors may often be accounted for in an adequate way. For example, if one assumes that these factors have comparable impacts on both mean and extreme high-water levels, an option is to first determine annual mean high-water levels, and subsequently, to consider only variations in extreme high-water levels relative to these annual means. Such an approach was taken, for instance, in von Storch and Reichardt (1997) who analyzed storm-related water level fluctuations at different tide gauges along the southern North Sea coast. An update of the analysis of von Storch and Reichardt (1997) is shown in Figure 4.12 (see color section). A strong and rather linear upward trend in annual mean high-water levels over about the last 160 years can be inferred. When annual extreme high-water levels are considered, a similar strong increase can be inferred (not shown). If we assume that, in a statistical sense, the storms will have a stronger impact on extreme than on mean water levels while local water works, mean sea level rise, etc. would have comparable impacts on both, the effects of the latter could be removed from the tide gauge data and the difference between annual extremes and annual mean water levels then represents a proxy for storm-related water level variations and finally for changes in the storm climate. A time series of such a proxy is shown in the upper curve of Figure 4.12. It can be inferred that the trend visible in annual mean high waters is no longer present. Instead the curve remained rather stationary over the past 150 years with some decadal fluctuations superimposed. The latter correlate well with the information obtained from storm activity indices based on high percentiles of geostrophic wind speed (see

Figure 4.11). Other options to determine storm proxies from tide gauge records are presented, for example, in Woodworth and Blackman (2002).

There is also a number of substantially different proxies using indirect evidence for changes in the storm climate. Essen *et al.* (1999) suggested an approach based on micro-seismic activity. This approach is motivated by the fact that micro-seismic records contain signals related to wave activity on the sea surface and thus to storminess. The inversion of, first, the embedding of wind statistics into the wave field and then into the micro-seismic activity was, however, found to be too complex to allow for reconstruction of homogeneous long-term records representative of well-defined regions. Other approaches use geochemical evidence that indicatea the intensity of upwelling in some areas (e.g., Black *et al.*, 2007; McGregor *et al.*, 2007), which in turn represents a measure of mean wind conditions. Sailing times on routes regularly served with sailing ships (García *et al.*, 2000) or costs for repairing dike damages (De Kraker, 1999) represent other sources of such indirect information.

4.4 GLOBAL REANALYSES AND REGIONAL RECONSTRUCTIONS

The state of the atmosphere, and of any other component of the Earth system, cannot be observed in its entirety. The various land-, ship-, aircraft-, and satellite-based observing systems provide point observations or vertical profiles distributed irregularly in space and time. These data are then used by operational weather centers to construct, or *analyze*, continuous distributions of atmospheric variables. According to Glickman (2000) an *analysis* refers to the procedure of projecting the state of the atmosphere or the ocean as known from a finite set of imperfect, irregularly distributed observations onto a regular grid, or to represent it by the amplitude of standard mathematical functions. As such, an analysis represents a space–time interpolation system and the data product derived from this procedure is usually also called an analysis.

Ideally, analyses incorporate all available measurements and our knowledge of the system as manifested in the numerical models used for the procedure. They are thus our best guess of the atmospheric or oceanic state at a given time. Because of our imperfect knowledge and the limited number of observations, analyses always deviate to some extent from the true but unknown state. Generally, larger scales are better described, simply because they are better sampled while smaller scales are subject to significant uncertainty.

Analyses were originally introduced to provide initial fields for numerical weather prediction. The models used for the procedure and their resolution have frequently been updated as knowledge increased and more powerful computers became available. While this was wanted and did not pose any problem for numerical weather prediction, it made analyses inhomogeneous and unsuited for climate studies; especially, for climate change studies. In particular, there are two types of inhomogeneities associated with analyses:

130   **How to determine long-term changes in marine climate?**   [Ch. 4

1. *Changes in the analysis system.* This comprises changes in the model or the data assimilation[13] system reflecting our advances in knowledge (such as upgrades in model parameterizations or numerics) as well as changes in model resolution including orography, exploiting the increasing computational power that became available in time.
2. *Changes in the observational network.* This includes changes in the density of the observational network, upgrade of measurement techniques, or changes in observational procedures. For example, before World War II the observational network was mostly based on surface observations. The radiosonde network that was subsequently built up was essentially confined to the extra-tropics of the Northern Hemisphere. In the Tropics and the Southern Hemisphere the upper-air network remained sparse until the advances of satellite data (Bengtsson *et al.*, 2004).

In recent decades major efforts have been undertaken to overcome these obstacles. These efforts are generally known as *reanalyses.* A reanalysis is similar to an analysis, except for two fundamental differences. First, it is not performed in real time,[14] and, second, it is performed with a *frozen* state-of-the-art numerical model and data assimilation system that do *not* change over the entire period for which the reanalysis is performed (Glickman, 2000). The data sets obtained from such reanalyses are not completely homogeneous. They are free from effects caused by *changes in the analysis system*, but not from effects caused by *changes in the observational network* (see above). In the following the effects of the latter will be discussed in some more detail. Subsequently, an overview about existing reanalysis products is given.

4.4.1   Can climate trends be estimated from reanalysis data?

The question was posed and investigated in a pioneering paper by Bengtsson *et al.* (2004). They computed several global quantities such as lower-tropospheric temperatures, integrated water vapor, or total kinetic energy from a specific—the ERA-40 (Uppala *et al.*, 2005)—reanalysis, and from observations. Subsequently, they explored the extent to which long-term trends in these quantities could be estimated from reanalysis data. To do so, they computed trends for different reanalysis periods; namely, before and after 1979 (i.e., before and after satellite data were assimilated).

For lower-tropospheric temperatures, Bengtsson *et al.* (2004) found that for the pre-satellite era the reanalysis data show a warming trend that is slightly stronger but still within error bounds when compared with the trend derived directly from satellite data. It is also about comparable with the trend obtained from sea surface temperatures (SSTs) for the same period. When the entire reanalysis period 1958–2001 was

---

[13] The process of combining diverse data sampled at different times, intervals, and locations, into a unified and consistent description of a physical system such as the state of the atmosphere (Glickman, 2000).

[14] In real time, data are processed immediately on their retrieval.

considered Bengtsson *et al.* (2004) found the trend from the reanalysis to be considerably stronger. Moreover, the trend turned out to be considerably larger than the SST trend for the same period. To elaborate on the consequences of the changes that occurred in the observing system in 1979, Bengtsson *et al.* (2004) performed a sensitivity experiment in which all satellite data were removed from the data assimilation, mimicking the situation before the advent of satellite observations. In this sensitivity experiment the lower-tropospheric temperature trend was largely reduced and the authors suggested that the ERA-40 data may have a systematic cold bias prior to the satellite era. Assuming that such a cold bias without data assimilation from satellites would occur for the entire reanalysis period, the introduction of satellite data into the assimilation system in 1979 may thus have caused an artificial warming trend when the period 1958–2001 is considered.

Kinetic energy can be determined realistically only from reanalysis data because of the sparsity of the radiosonde network and the difficulties in determining it globally and systematically from space observations. Bengtsson *et al.* (2004) computed annually averaged total kinetic energy from the ERA-40 reanalysis and found that the values tend to fall into two groups (Figure 4.13). Smaller values generally occurred before 1979 and larger values thereafter. No significant trends were found when the periods 1958–1978 and 1979–2001 were considered independently. Bengtsson *et al.* (2004) concluded that the trend over the entire reanalysis period is again mainly caused by changes in the global observing network and found that this was also supported by their sensitivity experiment in which satellite data were removed from the data assimilation. In this experiment, kinetic energy values before and after 1979 were about equal. Regionally, largest changes were found for the Southern Hemisphere extra-tropics, while changes for the Northern Hemisphere were smaller but still significant.

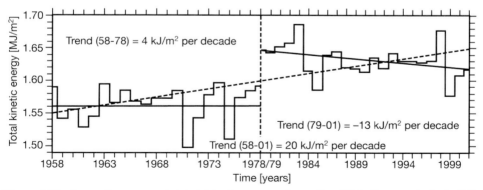

Figure 4.13. Annually averaged total kinetic energy from the ERA-40 reanalysis (solid), corresponding linear trends 1958–1978 and 1979–2001 (straight solid lines), and corresponding trend 1959–2001 (dashed). Redrawn from Bengtsson *et al.* (2004). Published/copyright (2004) American Geophysical Union. Reproduced/modified by permission of American Geophysical Union.

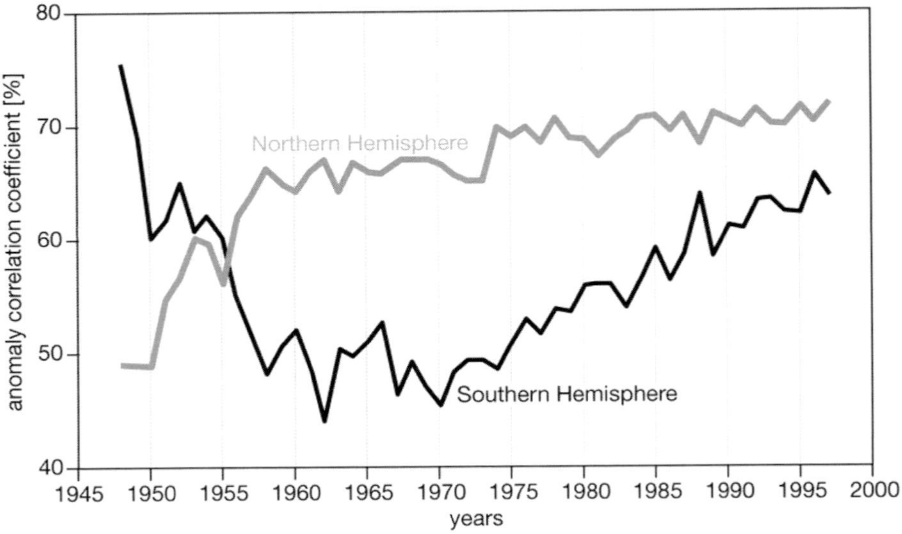

132   **How to determine long-term changes in marine climate?**   [Ch. 4

Figure 4.14. Annually averaged anomaly correlation coefficients between 5-day forecasts and reanalysis data at the time of the forecast, for the Northern Hemisphere (gray) and for the Southern Hemisphere (black). Redrawn after Kistler *et al.* (2001). Reproduced/modified by permission of American Meteorological Society.

Apart from the major, almost sudden change in the observing system in 1979, smaller and minor changes distributed over the course of time may also have introduced creeping inhomogeneities that may be harder to detect. This is illustrated in Figure 4.14 where development of the forecast skill is shown, when forecasts are initialized with reanalysis data. Here the forecast skill is expressed in terms of anomaly correlations; that is, the spatial correlation coefficient between the spatially distributed 5-day forecast and spatially distributed reanalysis data at the time of the forecast. Obviously, forecast skill is not constant, but increases steadily with time. This indicates that the quality, or the degree of realism, of the reanalysis has steadily improved due to better and more observations. Not surprisingly, reanalysis is much better in the Northern Hemisphere, where many more observations are available. Also, the level of homogeneity appears to be higher for the Northern Hemisphere. For the Southern Hemisphere, the high skill scores at the beginning of the reanalysis period prior to the International Geophysical Year 1957 are artificial. They are due to the fact that hardly any data were available in the Southern Hemisphere at that time with the result that reanalyses, with which forecasts were compared, were essentially the forecasts themselves.

When reanalysis data are used for the assessment of long-term changes in wind, wave, and storm surge climate the quality and reliability of reproducing tropical and extra-tropical cyclones becomes important. For extra-tropical cyclones Bromwich *et al.* (2007) investigated the extent to which their frequency and intensity distribution were affected by major changes in the global observing system in 1979. In particular,

Sec. 4.4] Global reanalyses and regional reconstructions 133

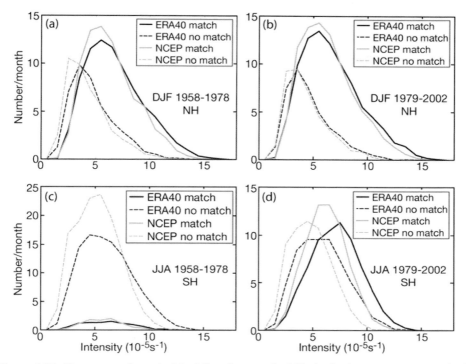

Figure 4.15. Frequency of matched (solid) and unmatched (dashed) winter storm systems in the Northern Hemisphere (upper row) and in the Southern Hemisphere (lower row) between the NCEP (gray) and the ERA-40 (black) reanalyses as a function of cyclone intensity given by the 850 hPa relative vorticity in 10^{-5} s^{-1}; analysis for 1958–1978 (left column) and for 1979–2002 (right column). Redrawn from Bromwich et al. (2007). Published/copyright (2007) American Geophysical Union. Reproduced/modified by permission of American Geophysical Union.

they compared the intensity and number of matched winter storm systems[15] in both hemispheres before and after 1979 for different global reanalyses (Figure 4.15). For the Northern Hemisphere, they found little change in cyclone intensity between 1958 and 2002 and only small differences between different reanalyses. Moreover, it was only weaker storm systems that did not match between different reanalyses. For the Southern Hemisphere, considerable differences in cyclone intensity before and after 1979 were analyzed. There are also very few cyclone matches before 1979 between different reanalyses suggesting that tracked storm systems are basically related with model climatology. After 1979 the number of matched systems increases significantly, although it remains low and is about comparable with the number of unmatched storms. It can be concluded that assimilation of satellite data is very important for the Southern Hemisphere and less so for the Northern Hemisphere, where a more developed ground-based observational system is in operation. Summarizing,

---

[15] That is, those systems that occurred simultaneously in different reanalyses.

134 How to determine long-term changes in marine climate? [Ch. 4

Bromwich *et al.* (2007) found very little change in Northern Hemisphere skill while changes and inhomogeneities are much more pronounced in the Southern Hemisphere. This suggests that reanalyses may provide a valuable tool and data base for the assessment of changes in the marine storm climate, at least for extra-tropical regions in the Northern Hemisphere.

The result is further supported by comparisons of proxy-based storm indices with that derived from reanalysis data. Paciorek *et al.* (2002) computed storm indices from reanalysis data based on geostrophic wind speeds following the approach presented in Alexandersson *et al.* (1998) and WASA-Group (1998) and compared the trends with that obtained from observations. They suggested that any data inhomogeneity in the reanalysis should manifest itself by different trends in storm indices derived from both data sets. For the region considered,[16] the results did not suggest that data inhomogeneity in the reanalysis plays a crucial role although some systematic bias in absolute values exists. A similar conclusion can be obtained from Weisse *et al.* (2009) who presented a comparison of different storm indices in Lund (Sweden) based on an approach presented in Bärring and von Storch (2004). The comparison between indices derived from observations and from reanalysis revealed systematic biases for some of the indices, but no discernible difference in long-term trends and inter-annual variability.

These findings are somewhat in conflict with the analyses presented in Smits *et al.* (2005), who compared trends in the annual number of different strong wind incidences over The Netherlands for the period 1962–2002. The statistics derived from a dataset of near-surface wind observations at Dutch meteorological stations indicated a decrease in storm activity over The Netherlands between 5% and 10% per decade while different reanalyses showed increasing storm activity over the period. Smits *et al.* (2005) associated these differences with inhomogeneities in the reanalysis data although inhomogeneity in the station data need to be considered as well.

The question on whether or not climate trends can be reliably estimated from reanalysis data thus strongly depends on the quantities and the regions considered. The problem may be negligible for time-averaged quantities with smooth spatial distributions such as monthly mean air pressure in well-observed regions (such as Europe where these quantities are well sampled all the time). For time series of extreme events—in particular, in data-sparse areas such as the Southern Hemisphere—the homogeneity problem may become severe and reanalyses may not represent a suitable tool for estimation of climate trends. This is particularly so for "unobservable" quantities, which are not based on direct observations, but which are completely determined by the model used in the reanalysis process. Such variables comprise, for instance, kinetic energy, surface fluxes, or precipitation rates.

Despite the deficiencies mentioned above, the reanalyses provide an optimal way to accurately interpolate data in space and time and a better way to obtain dynamical consistency between different parameters, even for those for which no direct measurements exist. If carefully applied, reanalysis data may very well be used to assess long-term changes and trends; in particular, in regions with a dense observa-

---

[16] Northern Europe and the northeast Atlantic.

Sec. 4.4]   **Global reanalyses and regional reconstructions**   135

tional network that has undergone only minor modifications. In order to make reanalyses more suitable for the study of climate trends one suggestion was to perform century-long reanalyses mimicking the state of the observing system at the beginning of the 20th century (Compo *et al.*, 2006). While for modern times the result of such an effort will not be comparable with the quality of existing reanalyses, more reliable estimates of long-term changes and trends may be expected as effects introduced by major changes in the global observing system are removed from the data.

4.4.2 The NCEP/NCAR Global Reanalysis

One of the longest and most up-to-date global reanalyses so far is the NCEP/ NCAR[17] reanalysis. In the following this reanalysis project will be discussed in more detail following the presentations by Kalnay *et al.* (1996) and Kistler *et al.* (2001).

The NCEP/NCAR reanalysis presently[18] covers the period 1948–2008 and provides a 60-year-long record of global atmospheric reanalyzed data. Today, this reanalysis is continuously updated, making it one of the longest, most up-to-date, and thus most valuable data sets for studies on long-term changes and climate variability. As a starting point, data from weather stations, ships, radiosondes, pibals,[19] aircraft, satellites, and other observational platforms are collected and quality controlled. Specifically, the following products are used:

— upper-air rawinsonde observations of temperature, horizontal wind speed components, and specific humidity,
— operational vertical temperature soundings over the ocean from polar-orbiting satellites,
— temperature soundings over land,
— cloud-tracked winds from geostationary satellites and surface marine wind speeds from polar-orbiting satellites,
— aircraft observations of wind and temperature,
— land surface reports of surface pressure, and
— oceanic reports of surface pressure, temperature, horizontal wind, and specific humidity.

The number of these data has considerably increased since the late 1940s. In particular, satellite data were routinely included after 1979. Thus, the quality of reanalysis in the early years cannot be as good as for more recent years. In particular, reanalysis of the earliest decade 1948–1957, when observations were few and primarily in the Northern Hemisphere, is less reliable than that of the last four decades.

The quality controlled data are subsequently assimilated into a state-of-the-art global spectral weather prediction model. Here the NCEP operational global weather

[17] National Centers for Environmental Prediction/National Center for Atmospheric Research.
[18] In early 2009.
[19] Pilot balloons—that is, balloons tracked visually or with radar.

prediction model as implemented on January 10, 1995 was used. Spatial resolution was reduced to T62 (see Section 3.3.2), equivalent to a horizontal grid spacing of about 210 km. In the vertical, 28 levels were used (Kalnay *et al.*, 1996). The numerical model is kept unchanged (frozen) and the system continues to be used with current data, so that reanalyses are available from 1948 to the present.

The result of this effort is a relatively homogeneous complete history of global weather maps. The maps have a horizontal gridding of about 210 km (T62 spectral truncation) and 28 pressure levels in the vertical. The assimilation method used the forecast model as a weak constraint; that is, the model is considered imperfect and its predictions are deemed uncertain. The resulting maps are thus not a solution of the model and they do not necessarily conserve quantities like mass, momentum, and energy. This has, however, turned out not to be a severe problem.

Reanalyses generate many variables. Some of them depend strongly on observational data while others depend more strongly on the forecast model and its parameterizations. To discriminate among the different classes of reanalysis output variables, four categories (A–D) were introduced:

A: For Type A variables, the analysis variable is strongly influenced by observed data.
B: For Type B variables, there are *some* observations that directly influence the value of the variable; however, the model also has a very strong influence on the value of the variable.
C: For Type C variables, there are no observations that influence the value of the variable and the latter is based solely on the model fields.
D: For Type D variables, the reanalysis represents fixed climatologies and is independent of model and assimilated observations.

Obviously, reliability is different for the different types of variables, with Type A variables being the most reliable ones. To get an impression of the quality and reliability of different reanalysis output variables Table 4.1 provides an overview of the most frequently used parameters. A complete listing can be found in Kalnay *et al.* (1996).

As numerical models and data assimilation techniques improve over time and additional data may become available, it may be useful to repeat the reanalysis procedure from time to time. For the NCEP/NCAR reanalysis such an effort has been provided by NCEP/DOE[20] (Kanamitsu *et al.*, 2002). This so-called Reanalysis 2 is presently available from 1979 onwards, fixes some errors, and updates parameterizations of physical processes in the NCEP/NCAR Reanalysis 1. It is essential to be aware of the existence of these two reanalyses and the differences between them, as mixing both products may introduce spurious and artificial signals when long-term changes are considered. Generally, it is believed that the upgraded Reanalysis 2 provides a better description of the state of the atmosphere, although this is not necessarily the case for all variables. For example, Winterfeldt (2008) analyzed

---

[20] National Centers for Environmental Prediction/Department of Energy.

Sec. 4.4] **Global reanalyses and regional reconstructions** 137

Table 4.1. NCEP/NCAR reanalysis output variables and category.

Variable	Units	Class	Type
Geopotential height	gpm	A	3D field
Horizontal wind components	$m\,s^{-1}$	A	3D field
Temperature	K	A	3D field
Absolute vorticity	s^{-1}	A	3D field
Pressure vertical velocity	$Pa\,s^{-1}$	B	3D field
Relative humidity	%	B	3D field
Temperature at tropopause	K	A	2D field
Pressure at tropopause	Pa	A	2D field
Horizontal wind speed components at tropopause	$m\,s^{-1}$	A	2D field
Vertical speed shear at tropopause	s^{-1}	A	2D field
Temperature at maximum wind level	K	A	2D field
Pressure at maximum wind level	Pa	A	2D field
Horizontal wind speed components at the maximum wind level	$m\,s^{-1}$	A	2D field
Pressure reduced to mean sea level	Pa	A	2D field
Pressure at the surface	Pa	A	2D field
Precipitable water	$kg\,m^{-2}$	B	2D field
Relative humidity of total atmospheric column	%	B	2D field
Potential temperature at the lowest sigma level	K	B	2D field
Temperature at the lowest sigma level	K	B	2D field
Pressure vertical velocity at the lowest sigma level	$Pa\,s^{-1}$	B	2D field
Relative humidity at the lowest sigma level	%	B	2D field
Horizontal wind speed components at the lowest sigma level	$m\,s^{-1}$	B	2D field
Maximum temperature at 2 m	K	B	2D field
Minimum temperature at 2 m	K	B	2D field
Specific humidity at 2 m	$kg\,kg^{-1}$	B	2D field
Temperature at 2 m	K	B	2D field
Zonal component of momentum flux	$N\,m^{-2}$	B	2D field
Horizontal wind speed components at 10 m	$m\,s^{-1}$	B	2D field

(*continued*)

138 **How to determine long-term changes in marine climate?** [Ch. 4

Table 4.1 (*cont.*)

Variable	Units	Class	Type
Cloud forcing net longwave flux at the top of atmosphere	W m^{-2}	C	2D field
Cloud forcing net longwave flux at the surface	W m^{-2}	C	2D field
Cloud forcing net longwave flux for total atmospheric column	Wm^{-2}	C	2D field
Cloud forcing net solar flux at the top of the atmosphere	W m^{-2}	C	2D field
Cloud forcing net solar flux at the surface	W m^{-2}	C	2D field
Cloud forcing net solar flux for total atmospheric column	W m^{-2}	C	2D field
Convective precipitation rate	$\text{kg m}^{-2}\,\text{s}^{-1}$	C	2D field
Clear sky downward longwave flux at the surface	W m^{-2}	C	2D field
Clear sky downward solar flux at the surface	W m^{-2}	C	2D field
Clear sky upward longwave flux at the top of the atmosphere	W m^{-2}	C	2D field
Clear sky upward solar flux at the top of atmosphere	W m^{-2}	C	2D field
Clear sky upward solar flux at the surface	W m^{-2}	C	2D field
Cloud work function	J kg^{-1}	C	2D field
Downward longwave radiation flux at the surface	W m^{-2}	C	2D field
Downward solar radiation flux at the top of the atmosphere	W m^{-2}	C	2D field
Downward solar radiation flux at the surface	W m^{-2}	C	2D field
Ground heat flux	W m^{-2}	C	2D field
Latent heat flux	W m^{-2}	C	2D field
Near-IR beam downward solar flux at the surface	W m^{-2}	C	2D field
Near-IR diffuse downward solar flux at the surface	W m^{-2}	C	2D field
Potential evaporation rate	W m^{-2}	C	2D field
Precipitation rate	$\text{kg m}^{-2}\,\text{s}^{-1}$	C	2D field
Pressure at high-cloud top	Pa	C	2D field
Pressure at high-cloud base	Pa	C	2D field
Pressure at middle-cloud top	Pa	C	2D field
Pressure at middle-cloud base	Pa	C	2D field
Pressure at low-cloud top	Pa	C	2D field
Pressure at low-cloud base	Pa	C	2D field
Pressure at the surface	Pa	C	2D field

Variable	Units	Class	Type
Run-off	$\mathrm{kg\,m^{-2}}/(6\,\mathrm{h})$	C	2D field
Nearby model level of high-cloud top	Integer	C	2D field
Nearby model level of high-cloud base	Integer	C	2D field
Nearby model level of middle-cloud top	Integer	C	2D field
Nearby model level of middle-cloud base	Integer	C	2D field
Nearby model level of low-cloud top	Integer	C	2D field
Nearby model level of low-cloud base	Integer	C	2D field
Sensible heat flux	$\mathrm{W\,m^{-2}}$	C	2D field
Soil moisture content	Fraction	C	2 layers
Total cloud cover of high-cloud layer	%	C	2D field
Total cloud cover of middle-cloud layer	%	C	2D field
Total cloud cover of low-cloud layer	%	C	2D field
Temperature of the soil layer	K	C	3 layers
Temperature of high-cloud top	K	C	2D field
Temperature of low-cloud top	K	C	2D field
Temperature of middle-cloud top	K	C	2D field
Zonal gravity wave stress	$\mathrm{N\,m^{-2}}$	C	2D field
Upward longwave radiation flux at the top of the atmosphere	$\mathrm{W\,m^{-2}}$	C	2D field
Upward longwave radiation flux at the surface	$\mathrm{W\,m^{-2}}$	C	2D field
Upward solar radiation flux at the top of the atmosphere	$\mathrm{W\,m^{-2}}$	C	2D field
Upward solar radiation flux at the surface	$\mathrm{W\,m^{-2}}$	C	2D field
Meridional gravity wave stress	$\mathrm{N\,m^{-2}}$	C	2D field
Visible beam downward solar flux at the surface	$\mathrm{W\,m^{-2}}$	C	2D field
Visible diffuse downward solar flux at the surface	$\mathrm{W\,m^{-2}}$	C	2D field
Meridional component of momentum flux	$\mathrm{N\,m^{-2}}$	C	2D field
Water equivalent of accumulated snow depth	$\mathrm{kg\,m^{-2}}$	C	2D field
Ice concentration	Integer	D	2D field
Land–sea distribution	Integer	D	2D field
Surface roughness	m	D	2D field

near-surface marine wind speeds over the northeast Atlantic and the North Sea from both reanalyses in comparison with observations. He found that for open sea, Reanalysis 1 shows a better agreement with observations, while Reanalysis 2 over-estimated mean wind speeds by up to $2\,\mathrm{m\,s^{-1}}$. Near the coast the situation is reversed with Reanalysis 2 showing better agreement with measurements. Winterfeldt (2008) concluded that this represents a highly implausible result. He showed that, with the exception of some subtropical areas, Reanalysis 2 generally showed higher near-surface wind speeds than Reanalysis 1. While differences between both reanalyses at upper levels appeared to be small, Winterfeldt (2008) found vertical wind profiles more reliable in Reanalysis 1 and concluded that near-surface marine wind speeds are biased on the high side in Reanalysis 2. It is therefore essential to note that for any study on long-term changes in the marine storm climate, Reanalysis 1 and Reanalysis 2 should not be mixed but clearly distinguished.

4.4.3 Other global reanalyses

There are a number of other global reanalysis projects carried out at different centers and institutes. At the European Center for Medium Range Weather Forecast (ECMWF) the first reanalysis was carried out in the early 1980s. Since then, two major reanalysis projects have been performed exploiting substantial advances in numerical models, data assimilation systems, and the observing system. ERA-15 (Gibson *et al.*, 1996) covered the period 1979–1993 and was completed in 1995. More recently, the ERA-40 project (Uppala *et al.*, 2005) covering the period 1957–2002 concluded. Compared with NCEP reanalyses both ECMWF reanalyses were performed at higher spatial resolution (Table 4.2). ERA-40 also featured an actively coupled wave model and a reanalysis of global sea states (e.g., Caires and Sterl, 2005). However, presently only NCEP reanalyses are continuously updated, while ECMWF employs a different strategy and is currently (in 2009) producing an interim re-analysis. This interim reanalysis uses a more advanced model and data assimilation system together with increased spatial resolution (T255, about 50 km) and focuses on reproduction of the data-rich period since 1989.

The Japan Meteorological Agency (JMA) has produced a reanalysis for the 25-year period 1979–2004 (JRA-25).[21] The spatial resolution of this reanalysis is comparable with that of ERA-15 (Table 4.2). JRA-25 is the first reanalysis to use wind profile retrievals surrounding tropical cyclones, making this reanalysis unique in the study of tropical storms. Other data that are only used by JRA-25 comprise, for example, SSM/I[22] snow coverage and digitized Chinese snow depth data.

There are a number of analysis/reanalysis efforts for the global ocean. They focus exclusively on ocean thermodynamics; that is, variables such as temperature, salinity, heat content, or sea level changes caused by thermal expansion. Storm surges and sea states are generally not considered. Comparison of nine major ocean analyses/reanalyses is presented in Carton and Santorelli (2008). Analyses differ in the choice

---

[21] *http://jra.kishou.go.jp*
[22] Special Sensor Microwave/Imager.

Table 4.2. Comparison of major atmospheric reanalysis projects.

Name, reference	Organization	Period	Resolution[a]	Assimilation[b]	Status[c]
JRA-25, http://jra.kishou.go.jp	JMA/CRIEPI	1979–2004	T106L40	3DVAR	Continues as JCDAS
ERA-15, Gibson et al. (1996)	ECMWF	1979–1993	T106L31	OI	Completed in 1996
ERA-40, Uppala et al. (2005)	ECMWF	1957–2002	TL159L60	3DVAR	Completed in 2003
NCEP/NCAR Reanalysis 1, Kalnay et al. (1996)	NCEP/NCAR	1948–present	T62L28	3DVAR	Continues as CDAS
NCEP/DOE Reanalysis 2, Kanamitsu et al. (2002)	NCEP/DOE	1979–present	T62L28	3DVAR	Continues as CDAS

[a] T: triangular truncation; TL: triangular with linear reduced Gaussian grid; L: vertical layers. T106 and TL159 are comparable with a latitude–longitude grid of about 110 km, T62 is comparable to about 210 km.
[b] 3DVAR: three-dimensional variational data assimilation; OI: optimal interpolation.
[c] CDAS: Climate Data Assimilation System.

of initial conditions, surface forcing (NCEP Reanalysis 1, NCEP Reanalysis 2, ERA-40), model parameterizations, and resolution. Moreover, observations used for production of the various analyses also differ (Table 4.3). Analyses mainly agree on large-scale features such as long-term variations in global ocean heat content. In particular, all analyses do show two periods of rapid warming, first in the early

Table 4.3. Comparison of major ocean analysis and reanalysis projects.

Name, reference	Period	Surface forcing	Analysis procedure
CERFACS, Davey (2005)	1962–2001	ERA-40	Sequential
GECCO, Köhl and Stammer (2008)	1950–1999	NCEP Reanalysis 1	4DVAR
GFDL, Sun et al. (2007)	1950–1999	NCEP Reanalysis 1	Sequential
GODAS, Behringer (2005)	1979–2005	NCEP Reanalysis 2	Sequential
INGV, Bellucci et al. (2007)	1962–2001	ERA-40	Sequential
ISHII, Ishii et al. (2006)	1955–2005	None	Objective analysis
LEVITUS, Levitus et al. (2005)	1955–2003	None	Objective analysis
SODA, Carton and Giese (2008)	1958–2005	ERA-40	Sequential
UKFOAM, Bell et al. (2004)	1962–1998	Met. Office	Sequential

142 **How to determine long-term changes in marine climate?** [Ch. 4

1970s and later beginning in the late 1980s (Carton and Santorelli, 2008). As for global atmospheric reanalyses, discrepancies are most pronounced in the Southern Hemisphere and at smaller spatial and time scales.

For storm surges so far no global reanalysis data set is known. Storm surges are usually modeled regionally with a specific interest for a particular coastal area. Such efforts are discussed in detail in Section 4.4.4. Sea states have been produced as part of the ERA-40 reanalysis where a global wave model was actively coupled to the atmosphere model. The results and quality of this wave hindcast were explored, for instance, in Sterl and Caires (2005), Caires and Sterl (2005), and Caires *et al.* (2005). In all other global atmosphere reanalyses no wave model was included. However, their wind fields have subsequently been used to produce global sea state hindcasts with standalone ocean wave models. For ERA-15, results can be found in Sterl *et al.* (1988), for the NCEP/NCAR Reanalysis 1 they are described in Cox and Swail (2001). There are also a number of regional efforts using either reanalysis or downscaled reanalysis winds. They are discussed in Section 4.4.4.

4.4.4 Regional reanalyses and reconstructions

Global reanalyses are technically and computationally demanding endeavors. Because of this, the space and time resolution of global reanalyses are generally limited and the number of observations that can be taken into account is restricted. It appears obvious to address and to reduce these restrictions by regionally constrained and better resolved approaches. In principle, there are two different strategies:

1. *Regional reanalyses* use the same approach as global reanalyses. In particular, a regional model and a data assimilation system at improved resolution are implemented. At the lateral boundaries they are driven by global reanalysis data and in the interior of the model domain additional data may be assimilated that may or may not have been accounted for in production of the global reanalysis. Due to the limited model domain, increased spatial resolution is possible. In general, however, the approach remains complex and time-consuming.
2. *Regional reconstructions* feature a different, more simple, and less expensive strategy. They are based on the idea that the skill of global reanalyses is scale-dependent with higher/lower skills for larger/smaller scales. As for regional reanalyses, a regional model is implemented that is driven by global reanalysis data at its boundaries. In the interior of the model domain, however, no additional data are assimilated. Instead, techniques are applied that keep the regional model solution close to that of the global reanalysis for larger scales that are well supported by data assimilation in the global reanalysis. At the same time these techniques allow the regional model to evolve independently from the global reanalysis for smaller scales that have little skill in global reanalysis data. Examples of such techniques include, for example, spectral nudging (von Storch *et al.*, 2000) (see also Section 3.3.3) or scale selective bias correction (Kanamaru and Kanamitsu, 2006). The approach is sometimes also referred

to as regional data assimilation without observations (von Storch *et al.*, 2000). Note that the term *regional reconstruction* is not well established in the scientific literature. It is used here to distinguish the approach from regional reanalyses.

An example of a regional reanalysis is the so-called North American Regional Reanalysis (NARR) (Mesinger *et al.*, 2006). Here a regional model and data assimilation system, set up for North and Central America, were driven by the NCEP/DOE Global Reanalysis 2 from 1979 onwards. Compared with the driving global reanalysis, the regional system employs a largely increased spatial resolution with 32 km horizontal grid spacing and 45 levels in the vertical. Additionally, in the interior of the domain, data are assimilated that have not been considered in the global reanalysis. These comprise, for example, precipitation, near-surface winds, sea and lake ice, or temperatures, especially for the Great Lakes. Mesinger *et al.* (2006) showed that compared with the driving global reanalysis considerable improvements in reanalysis skill could be achieved for many variables. They concluded these improvements to result from a combination of better/more observational data in the regional reanalysis, of a better assimilation scheme, improved model parameterizations, and higher spatial resolution.

One of the first attempts to provide a multi-decadal regional atmospheric reconstruction was provided by Feser *et al.* (2001). The attempt mainly relies on the spectral nudging technique developed in von Storch *et al.* (2000) and uses a regional atmosphere model at about 50 km spatial resolution driven by the NCEP/NCAR Global Reanalysis 1 to reconstruct atmospheric conditions for Europe over the past decades. Feser (2006) and Winterfeldt (2008) demonstrated that, compared with the driving global reanalysis, the approach indeed produced better results for many variables; in particular, on medium to small spatial scales that are not well supported in global reanalysis. The data set initially covered the period from 1958 to 1998. It was continuously updated and is now, in early 2009, available from 1948 to 2007.

A similar approach is described by Kanamitsu and Kanamaru (2007) who presented a regional atmospheric reconstruction for California. They used a regional atmosphere model at 10 km resolution implemented over California and driven by the NCEP/NCAR Global Reanalysis 1 to reconstruct atmospheric conditions for 1948 to 2005. Here a scale-selective bias correction technique (Kanamaru and Kanamitsu, 2007) was applied to keep the regional model solution close to the driving global reanalysis for large scales. The data set is usually referred to as California Reanalysis Downscaling at 10 km (CaRD10). Kanamitsu and Kanamaru (2007) showed that, compared with the driving global reanalysis, improved skills especially for near-surface wind fields could be obtained this way. Comparing CaRD10 with NARR, Kanamaru and Kanamitsu (2007) found higher skills in CaRD10 and concluded that this may be a result of either the further increased spatial resolution in CaRD10 or that near-surface observations are not well assimilated in current state-of-the-art data assimilation systems.

For storm surges and wind waves no regional reanalyses are available. Instead, regional reconstructions (or hindcasts) are provided that use atmospheric forcings either from global reanalyses or from regional reanalyses or reconstructions. For storm surges, numerous such applications do exist. The first attempt to reconstruct the storm surge climate over decades of years was provided within the WASA project (WASA-Group, 1998) by Flather *et al.* (1998) for the North Sea. Similar approaches were provided later by, for example, Langenberg *et al.* (1999), or more recently by Weisse and Pluess (2006) who used improved models, increased spatial and temporal resolution, and improved atmospheric forcings to reconstruct longer periods. Similar approaches exist for other areas as well, such as for the Mediterranean Sea (Ratsimandresy *et al.*, 2008), the Baltic Sea (e.g., Jedrasik *et al.*, 2008; Meier *et al.*, 2004), and others. For waves, one of the first attempts to reconstruct many decades was provided by Günther *et al.* (1998) for the North Sea. More recently, Weisse and Günther (2007) used improved meteorological wind fields and a wave model at higher spatial resolution to reconstruct wave conditions in the North Sea over the period 1948–2007. Similar efforts were made for the eastern Mediterranean Sea (Musić and Nicković, 2008), the Baltic Sea (Cieślikiewicz and Paplińska-Swerpel, 2008), and the waters around New Zealand (Gorman *et al.*, 2003).

So far, all regional reanalyses/reconstructions have focused solely either on atmospheric or on marine aspects; in particular, waves and storm surges. More recently, for the first time, an effort was made to provide a consistent meteorological marine reconstruction of atmospheric and marine conditions. Here a series of models was used. First, the NCEP/NCAR Global Reanalysis 1 in combination with a spectral nudging approach (see Section 3.3.3) was used to drive a regional atmosphere model over Europe with a spatial grid size of about 50 km. From this atmospheric reconstruction, near-surface marine wind fields were subsequently used to drive high-resolution wave and tide–surge models. While the wave model was run in a nested mode with a coarser grid (about 50 km grid size) covering most of the northeast Atlantic and a finer grid (about 5 km grid size) covering the North Sea south of 56°N, the tide–surge model was run on an unstructured grid with typical grid spacing of about 5 km in the open North Sea and largely increased values (up to 80 m) near the coast and in the estuaries. The system was used to reconstruct the period 1948–2007 and is now continuously updated. Output from all models is stored every hour. In this way a high-resolution meteorological marine data set for the North Sea covering the last six decades was created. An overview of this reconstruction and its application to a number of problems ranging from offshore wind and ship design to coastal protection and the assessment of chronic oil pollution is provided in Weisse *et al.* (2009). The data set is known as *coastDat*[23] and is continuously extended to include additional areas and parameters.

4.5 REGIONALIZATION TECHNIQUES

Due to computational constraints, global reanalyses or scenarios (see Section 4.6) often have too coarse a spatial resolution to be used for regional and local impact or

---

[23] See *http://www.coastdat.de*

climate change studies. Therefore, a number of regionalization techniques have been developed with the objective of enhancing regional information. Most of these techniques have been briefly mentioned in previous sections. Here, a more systematic comparison and discussion are provided.

There are, in principle, three different approaches to regionalization.

1. *High-resolution global models* may be used in some cases when sufficient computational power is available. In particular, global models with variable resolution are applied where denser grid spacing in the region of interest is used.
2. *Regional models* with higher spatial resolution may be implemented over the region of interest and nested within global models to obtain enhanced spatial resolution at the regional scale. The models are usually driven by data from global models at lateral boundaries. Additionally, techniques such as spectral nudging or scale-selective bias correction may be applied in cases where the large-scale features of global simulation are supported by data assimilation (see Sections 3.3.3 and 4.4.4 for details).
3. *Statistical methods* are an option when some observational data on the regional scale or some results from regional models are already available and when only limited computational power can or should be used.

The latter two approaches are usually referred to as *downscaling*,[24] the rationale for which was developed in Section 1.4.2. It is based on the premise that regional climate is to some extent controlled by large-scale climate. Under this assumption, the variability of regional climate can be described as a superposition of one part that is controlled by and another that is uncontrolled by large-scale climate—see Equation (1.9). The contribution of both terms to overall regional climate variability may vary. Generally, the larger the part controlled by large-scale climate—that is, the stronger the physical link between large-scale and regional-scale climate variables—the more successful both downscaling approaches may be. Implicitly, all downscaling studies assume that the link is strong and that that part of regional variability that is not controlled by large-scale climate is small.

Regionalization techniques that utilize regional models are also referred to as *dynamical downscaling*. The technique essentially originated from numerical weather prediction and its use for climate applications was pioneered by Dickinson *et al.* (1989) and Giorgi (1990). To date, it is widely used for climate and climate impact studies with typical spatial grid sizes of a few (tens of) kilometers in the case of regional atmospheric models. For waves and storm surges, grid sizes may vary considerably depending on region, model, and type of applications. Often, a few kilometers are typical (e.g., Langenberg *et al.*, 1999; Weisse and Günther, 2007); however, in some cases a few hundred meters have already been used (e.g., Weisse and Pluess, 2006). Generally, the nesting of regional models is *one-way*; that is, there is no feedback from the regional-scale model on the large-scale model and scales. In

---

[24] A term introduced by von Storch *et al.* (1991).

the framework of (1.9) this means that the first term on the right-hand side is ignored and regional climate variability is completely controlled by larger scales.

The same assumption is implicitly made when *statistical downscaling* methods are applied. Here, some observations representing large-scale (*predictors*) climate and regional-scale (*predictands*) climate are usually needed. Based on these observations, a statistical model is built between predictors and predictands, which may be used subsequently to analyze the impact of changes/variations in large-scale parameters on regional climate. A range of different statistical techniques has been developed ranging from regression and canonical correlation analysis to neural networks and analogs. For an overview we refer the reader to, for example, Zorita and von Storch (1999).

A special type of statistical downscaling is commonly referred to as *statistical–dynamical* downscaling. Here, a classification is first made for an often large number of weather situations (e.g., according to wind direction and stability). Then for each of the situations, a dynamical model is run and regional climate is described in spatial detail by weighting the resulting fields by the relative frequency of occurrence of weather classes. Also, the effect of climate change can be considered in this way.

All downscaling methods have advantages and disadvantages. Statistical approaches are computationally inexpensive and can be applied, for instance, to a large number of climate change experiments. When sufficiently good regional data are available, they can also provide local information in terms of variables that are of primary interest for impact studies. On the other hand, the technique only provides information on variables for which *a priori* information is available to construct the statistical model and for places at which such observations have been made. Dynamical downscaling techniques, in contrast, may provide a consistent, spatially and temporally resolved picture of regional climate, also of places and variables for which no observations have been made. Their main limitations are that they are computationally expensive and that their degree of realism depends on the quality of the models involved.

The third regionalization technique, which relies on the use of highly and variably resolved global models, differs significantly from the previously discussed downscaling techniques. In the framework of Equation (1.9), the first term on the right-hand side (i.e., feedbacks from regional scales on larger scales) are explicitly accounted for. The major disadvantages of this approach are that it is computationally even more expensive and that feedbacks from smaller scales on larger scales are represented only partially (i.e., only as generated within the region where enhanced spatial resolution was applied). As such global models with variable resolution may lead to improper description of small-scale to large-scale feedbacks (Giorgi *et al.*, 2001). The severity of the problem again depends on the region and the parameters of interest.

A fundamental question in the application of regionalization techniques—in particular, for dynamical downscaling and for application of variable resolution global models—is whether or not an *added value* can be obtained from application of these approaches. While it seems obvious to assume that increased spatial and/or time resolution provides enhanced regional information, it is not *a priori* clear

Sec. 4.5] Regionalization techniques 147

whether or not the skill in the representation of regional variables has increased by application of these techniques. We briefly addressed the issue in Section 3.3.3 and now return to this point in more detail.

One cannot expect regional modeling to improve all aspects of regional climate. Instead, certain aspects will be described better, while others will not. This in turn depends on the variables, scales, and regions considered. For example, García-Sotillo et al. (2005) examined the added value of dynamical downscaling the NCEP/NCAR Reanalysis 1 for the Mediterranean Sea and adjacent parts of the northeast Atlantic. In general, they found very little added value for sea level pressure when comparing NCEP/NCAR and downscaled reanalysis fields with observations. However, for temperatures at 2 m height considerable improvements in downscaled fields were obtained; in particular, when temperature extremes or coastal and inland stations surrounded by complex orography were considered. Similarly, for near-surface marine wind speeds, no improvements were obtained far off the coast, while there were noticeable enhancements for some coastal stations in the Mediterranean Sea. Again, García-Sotillo et al. (2005) noted the improvement to be more noticeable for extreme conditions. Figure 4.16 shows an example from the analysis of García-Sotillo et al. (2005). Here percentiles derived from wind speed time series recorded at two buoys were compared with percentiles derived from regional downscaling and from the driving NCEP reanalysis. For the Atlantic station well off the coast, reanalysis and downscaled wind speed percentiles are very similar both to each other and to

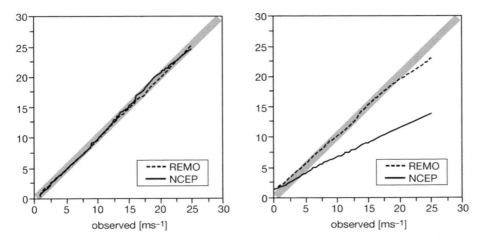

Figure 4.16. Validation of near-surface marine wind speeds at two buoy locations representative of offshore conditions in the Atlantic (left) and of near-shore conditions in the Mediterranean Sea (right). Shown are percentile–percentile plots of observed (x-axis) vs. NCEP-reanalyzed wind speeds (y-axis, solid), and vs. spectrally nudged and regionally downscaled NCEP wind speeds (y-axis, dashed). The expected range of uncertainty is indicated by the gray band. In the case of the Atlantic buoy observational data were also assimilated into the NCEP reanalysis; in the case of the Mediterranean buoy observational data were independent. Redrawn after García-Sotillo et al. (2005).

observed percentiles. For the coastal station in the Mediterranean Sea, only the dynamically downscaled wind field provided the right level of strong wind speeds. García-Sotillo *et al.* (2005) concluded that, apart from the variable considered, the added value depends on various geographical features such as type of basin, distance to the coast, or complexity of the orography surrounding the basin.

In the example provided in Figure 4.16, near-surface marine wind speeds from the offshore (Atlantic) location were assimilated into the NCEP reanalysis, while data from the coastal (Mediterranean) station were not. A more systematic approach to distinguish between the effects of data assimilation and distance to the coast on the added value of dynamically downscaled reanalyses was provided by Winterfeldt (2008). He compared near-surface marine wind speeds from a number of different offshore and near-shore locations and additionally distinguished between their assimilation status. He found that assimilation status had no significant impact on added value assessment and confirmed the conclusions of García-Sotillo *et al.* (2005) that for near-surface marine wind speeds the added value depends primarily on distance to the coast and, for near-shore locations, on complexity of the surrounding orography.

Winterfeldt and Weisse (2009) collocated near-surface marine wind fields from satellite data with NCEP reanalysis and regionally downscaled NCEP reanalysis wind fields over the area covered by the regional model (Europe and adjacent seas). Subsequently, they computed a skill score S defined by

$$S = 1 - \sigma_{fo}^2 \sigma_{ro}^{-2} \tag{4.1}$$

where σ_{fo}^2 denotes error variance between regional model data f and verifying satellite observations o; and σ_{ro}^2 is the corresponding error variance between NCEP reference data r and satellite observations. According to this definition, S measures the skill of regional model data f relative to reference data r of the same predictand o. When $S > 0$, then the winds downscaled by the regional model fit the observations (satellite data) better than NCEP reanalysis winds. In this case, added value or additional knowledge in comparison with the reanalysis was achieved. Figure 4.17 (see color section) shows the result of this exercise. It confirms the findings of García-Sotillo *et al.* (2005) and Winterfeldt (2008) obtained from comparison with *in situ* measurements; in particular, that there is no added value for near-surface marine wind speeds over the open ocean, and that near the coasts the added value depends on distance to the coast and complexity of the nearby orography and coastline.

Feser (2006) studied the scale dependence of the added value. In particular, she applied appropriate spatial filter techniques on several parameters to separate different scales. For spatially smooth (large-scale) variables such as near-surface air pressure, she found improvements limited to medium scales, while for regionally more structured (small-scale) quantities such as near-surface air temperatures, the added value was more obvious and present on all scales. Summarizing, the answer to the question on whether or not there is added value in the application of regionalization techniques strongly depends on the variable and the problem considered. Moreover, geographical aspects of the study region such as complexity of the

Sec. 4.6]   **Scenarios and projections**   149

orography or distance to the coast play a role. Regionalization techniques are thus not *per se* useful, but their design and application should be carefully planned depending on the problem considered.

4.6 SCENARIOS AND PROJECTIONS

Scenarios are a commonly used technique to assess the consequences of possible future developments and to elaborate on options to deal with them or to minimize their impact. For example, when a children's birthday party is planned we may prepare for two scenarios: an outdoor party in case of sunshine, or an indoor party in case of rainy weather. Both scenarios are plausible, possible, and internally consistent. In particular, they deal with the question *What* do we do *if* the weather turns out to be good or bad. Scenarios are thus fundamentally different from predictions, in which *unconditional* statements about the future are made; for example, "the weather will be rainy on the day of the birthday party."

Generally, scenarios are defined as descriptions of possible, plausible, and internally consistent, but not necessarily probable future states. In contrast to a prediction, the main purpose of a scenario is to provide a framework in which the availability and adequacy of options to respond to possible future conditions can be analyzed. Such options may include, for example, implementation of measures to avoid unwanted future states or development of options to adapt to and reduce the consequences of unwanted futures (von Storch, 2007).

The scenario technique is widely used in climate research; in particular, for assessing the consequences of anthropogenic climate change (*anthropogenic climate change scenarios*). In the following, we briefly discuss these scenarios and the approach, following mainly the discussion presented in von Storch (2007).

Anthropogenic climate change scenarios (hereafter shortly referred to as *climate change scenarios*) are built in a series of steps.

1. *Emission and concentration scenarios* are created first. Emission scenarios are socio-economic scenarios based on *ad hoc* assumptions that finally lead to emissions. Assumptions inherent in these scenarios comprise population growth, the efficiency of energy use, and future technological developments (Tol, 2007). Subsequently, emission scenarios are processed and transferred into scenarios of greenhouse gas concentrations.
2. *Climate change scenarios* are derived by forcing concentration scenarios on global climate models. As a result, possible, plausible, and internally consistent estimates of future climate conditions (means, variability, seasonal cycle, etc.) are obtained. Methodologically, this step represents a *conditional prediction*, as once the emission scenario has been defined no further *ad hoc* assumptions are required and each emission scenario corresponds to a specific climatic development. The process of obtaining climate change scenarios is sometimes referred to as *climate projection* although the phrase is not used consistently throughout the literature.

150   **How to determine long-term changes in marine climate?**   [Ch. 4

3. *Regional downscaling* (see Section 4.5) using regional climate, wind wave, storm surge, or other models may be applied to estimate the regional consequences of underlying emission scenarios. In this case, it is often advisable to consider at the same time the possibility that other relevant aspects may change. For example, a climate change–induced alteration in wind climate may alter storm surge heights in an estuary. At the same time, dredging of waterways or modifications to coastal protection may have similar effects on storm surge height. In order to elaborate consequences and to develop options in a consistent way, scenarios for regional and local social and economic developments are needed and have to be considered jointly with climate change projections (e.g., Bray *et al.*, 2003; Grossmann *et al.*, 2006).

Scenarios may in some cases become predictions; namely, when all sets of scenarios share a common feature (i.e., when conditional predictions agree independently of underlying assumptions: conditioning). For example, all climate change scenarios project an increase in global mean temperature and a rise in global mean sea level. Although, the magnitude of the increase depends on underlying assumptions (i.e., greenhouse gas emissions), the sign of the change does not. In other words, based on present knowledge, the unconditional prediction can be made that global mean temperature and sea level will rise towards the end of the 21st century.

As some understanding of the nature of emission scenarios is required for the assessment of results presented in Chapter 5, some basic features of these scenarios will be discussed in the following. A number of frequently used emission scenarios in climate research have been published in the IPCC[25] special report on emission scenarios (Nakicenovic and Swart, 2000). They are usually referred to as SRES scenarios. Depending on underlying socio-economic assumptions, SRES scenarios can be grouped into four main *families*. Within each family all scenarios share the same demographic, politico-societal, economic, and technological storyline.

A1: This family of scenarios describes a world of rapid economic growth and rapid introduction of new and more efficient technologies.
A2: This family describes a very heterogeneous world with emphasis on family values and local traditions.
B1: This family describes a world of "dematerialization" and introduction of clean technologies.
B2: This family describes a world with an emphasis on local solutions to economic and environmental sustainability.

Figure 4.18 (see color section) shows the emissions for carbon dioxide (a representative of greenhouse gases) and sulfate dioxide (a representative of anthropogenic aerosols) obtained for some scenarios in each family. It can be inferred that the time development of emissions is quite different in the different scenarios. For example, in the case of carbon dioxide, emissions in 2040 are considerably larger in the A1 than in the B2 scenarios. Later, as more efficient technologies are developed, emissions in the

---

[25] Intergovernmental Panel on Climate Change.

A1 family decrease significantly in most cases while a more or less steady increase is inferred for B2. Figure 4.18 also illustrates the process by which these emissions are transferred to global climate change projections. The emissions are transferred to concentrations, which for carbon dioxide show an upward trend towards the end of the century independent of the chosen emission scenario. Concentrations are then used to force global climate models from which projections of future temperature and sea level change are obtained. The latter is illustrated in the two panels at the bottom of Figure 4.18. As an increase in global mean temperature and sea level is obtained from all emission scenarios independent of underlying socio-economic assumptions, the increase in both variables towards the end of the 21st century becomes an unconditional prediction.

SRES scenarios are not unanimously accepted by the economic community. Some researchers find the scenarios internally inconsistent (Tol, 2007). Documentation of the various points raised is provided by the House of Lords, Select Committee of Economic Affairs.[26] A key critique is that the expectation of economic growth in different parts of the world is based on market exchange ranges and not on purchasing power parity (Kellow, 2007). Another aspect is the implicit assumption in SRES scenarios that the difference in income between developing and developed countries will significantly shrink by the end of this century (Tol, 2006). The argument is that these assumptions lead to an exaggeration of expected future emissions.

In the Fourth IPCC Assessment Report (Solomon *et al.*, 2007) uncertainty assessments are provided and, for some variables, probabilities are assigned to expected future changes. It should be emphasized that uncertainty assessment appears to be subjective and is not used consistently throughout the report. Also, assigning probabilities to projected future changes turns out to be problematic. The reason is that the assumption is made implicitly that the sample of scenarios and climate model simulations represents a random sample that represents the entire population; that is, the whole range of possible outcomes. This assumption is likely to be violated for both. For scenarios it appears likely that not the entire range of possible socio-economic developments is covered, and for climate change simulations one or more models may share the same deficiencies or model biases because many of the models have common roots. The uncertainty and probability assessments provided by the IPCC Fourth Assessment Report may thus, at best, be considered lower bounds.

4.7 DETECTION AND ATTRIBUTION

Climate variability appears to be a superposition of internally driven and externally forced fluctuations (see Section 1.4.1). Often, interest in the analysis of climate

---

[26] House of Lords, Select Committee on Economic Affairs, 2005: *The Economics of Climate Change*, Vol. I: *Report*, 2nd Report of Session 2005–06, Authority of the House of Lords, London; The Stationery Office Limited, HL Paper 12-I, *http://www.publications.parliament.uk/pa/ld/ldeconaf.htm\#evid.*

152   **How to determine long-term changes in marine climate?**   [Ch. 4

variability is in determining whether or not an observed change appears to be unusual in a statistical sense and whether some hypotheses about the causes can be established and tested. A framework for this kind of analysis was originally developed and introduced by Hasselmann (1979). Later the approach was developed and adapted towards the problem of anthropogenic climate change (Hasselmann, 1993). The framework is commonly designated as *detection and attribution.* Following the definitions provided in the Third and Fourth Assessment Reports of the IPCC (Houghton *et al.*, 2001; Solomon *et al.*, 2007), its ingredients are

1. *Detection* refers to the process of demonstrating that an observed change is, in a statistical sense, significantly different from what can be explained by natural internally driven variability. Here internally driven variability refers to variations and fluctuations that occur in the absence of anomalous external forcing (IDAG, 2005). Detection does not imply attribution of the cause of the detected change.
2. *Attribution* represents the process of demonstrating that a detected change is consistent with a hypothesis about the effects from a prescribed combination of external forcings (including anthropogenic ones) and not consistent with alternative and physically plausible hypotheses and explanations (IDAG, 2005).

In the following, the two components of detection and attribution are discussed in more detail.

4.7.1   The detection problem

Detection represents a statistical signal-to-noise problem. When anthropogenic climate change is considered, the challenge is to find, in an observational record, evidence for contamination of natural climate variability by forced signals. Formally, the null hypothesis is tested that the observed climate signal S_t is consistent with natural variability

$$S_t \in P[\mu_0, \sigma_0] \tag{4.2}$$

where t denotes time; and P describes the distribution of the present climate with a location parameter (mostly mean) μ_0 and a dispersion parameter (often standard deviation) σ_0. The problem is to define S_t properly and to determine μ_0 and σ_0. In the case of anthropogenic climate change, the detection variable S is the *trend* of a climate variable; for instance, global mean temperature, sea level, or large-scale precipitation. The question frequently raised is: *Is there a trend in the data that is "statistically significant"?* This question is not well posed. The underlying null hypothesis is implicitly made and tested that there is no trend in the data; that is, that a series of randomly drawn numbers have constant mean values (Zwiers and von Storch, 2004). Even if such a test has formally been performed correctly (for problems see Section 4.7.3), what does the presence of a "statistically significant trend" (i.e., a trend statistically significantly different from zero) imply? The meaning is that were the experiment repeated in the future, the expectation would be to find a trend again. For example, were a mid-latitude temperature time series from January to August screened for trends such a signal will be found because of the annual cycle. In a statistical sense this trend would be "significant", meaning that if we repeat the experiment under the same conditions (e.g., by repeating the trend analysis on January–August temperature data from the following year), we expect to see a similar trend again. However, this finding (the "significance" of the trend) provides no information on whether or not the trend is likely to continue into the future, whether (in our example) it continues if the time series is extended into September, October, November, and so forth. The problem here is that the mathematics behind the word "significance" are not adequately recognized and blended with the meaning of "significance" in everyday language (i.e., "relevance"). Thus, if there is a trend in the observational record, the relevant question is not whether it is significantly different from zero, but whether its slope is different from what would be expected from an undisturbed time series. This again represents a detection problem in which naturally occurring trends are being considered to be drawn from a random variable with a distribution function $F(p)$ of $P[\mu_0, \sigma_0]$. A trend which cannot be explained by natural variability alone is said to be "detected" when the hypothesis is rejected that recent trends S_t are sufficiently rare under the null hypothesis; that is, $S_t > F(95\%)$.

For estimation of the parameters of P different options exist. They may be estimated from the limited observational record, provided there is sufficient space–time coverage and homogeneity of the data, and the data set is free of marked contamination by external factors. An example is the study by Rybski *et al.* (2006) who used a variety of historical reconstructions of global mean temperature. Often, results from control simulations with quasi-realistic climate models are used, however (see Section 3.3.1).

4.7.2   The attribution problem

Attribution refers to finding the most plausible explanation for observed (detected) changes. In contrast to detection, it is less rigorous and mainly addresses plausibility issues. Formally, after signal S was found to be outside the present climate distribution P, a mix of processes (explanations) is linked (attributed) to the signal that fits best with *a priori* assumed links between causes and effects

$$S_t = \sum_k \gamma_k R_k + N_t, \tag{4.3}$$

where N_t denotes time-dependent noise (residuals); R_k are (time-dependent) response patterns of hypothesized (prescribed) external forcing functions, the effects of which are tested for; and γ_k represent a set of scaling factors.

Usually there are many external forcing factors. In the case of attributing anthropogenic climate change, climate theory limits their number to just a few. Among them are changing concentrations of greenhouse gases in the atmosphere, or the effects of volcanic aerosols and solar activity. The response patterns R of these forcings are usually obtained from experiments with quasi-realistic numerical climate models, in which the response of the system to (a set of) prescribed forcing factors is simulated. The response patterns R are usually represented as patterns of variation in space, time, or both and are commonly referred to as *fingerprints* (Solomon *et al.*, 2007).

Once fingerprints are designated, it is determined whether they are consistent with patterns detected in observations. Figure 4.19 (see color section) illustrates the procedure in the case of global mean temperature. It shows the scaling factors γ by which the amplitudes of several model-simulated signals R have to be multiplied to reproduce corresponding changes in the observational record. A range encompassing unity implies that this combination of forcing and model-simulated response is consistent with observed changes, while a range encompassing zero implies that this model-simulated signal is not detectable in observations (Mitchell *et al.*, 2001). The first entry on the left side shows the scaling factor and the 5%–95% uncertainty range obtained from one model forced only by changing greenhouse gas concentrations. The factor is considerably smaller than one, indicating that models forced only by changing greenhouse gas concentrations considerably overestimate the observed warming trend (Mitchell *et al.*, 2001). This result is consistent with that obtained from other models. The next eight entries show the scaling factors obtained from different models forced by greenhouse gases and aerosols (sulfate). All but one of these factors are consistent with unity; that is, the model-simulated responses are in agreement with observed warming. The latter implies that, according to the detection and attribution formalism introduced, the currently observed trend in global mean temperature can be attributed to a combination of changing greenhouse gas concentrations, changing aerosol concentrations, and internal natural variability of the climate system.

The detection and attribution framework is meanwhile well established in climate research; in particular, when anthropogenic climate change is considered. There are now many studies, concluding that a human influence on observed climate change on global to sub-continental scales does exist (e.g., Hegerl *et al.*, 1996; Stott, 2003). The magnitude of human influences on regional scales is still a matter of debate because of smaller signal-to-noise ratios. Recently, there have been some studies showing noticeable influences on smaller scales as well (e.g., Bhend and von Storch, 2007, 2009; Wang *et al.*, 2009).

4.7.3 Trends and detection

In the case of detecting anthropogenic signals in the observational climate record, the eminent detection variable is trends. In the following we briefly discuss some issues related to the detection of trends.

A most widely used statistical test to determine whether the appearance of a time series is consistent with the absence of a trend is the Mann–Kendall trend test (Mann, 1945; Kendall, 1970; Hamed, 2009). We first determine the "ranks" R_i of the observations x_i $(i = 1, \ldots, n)$ in the time series. Rank $R_i = 1$ is given to the smallest x_i, rank $R_i = 2$ to the second smallest, and so forth, and rank $R_i = n$ to the largest of the n observations x_i. Then

$$S = \sum_{i=1}^{n-1} \sum_{j=1}^{n} \text{sign}(R_j - R_i), \qquad (4.4)$$

with $\text{sign}(a) = -1$ if $a < 0$, $\text{sign}(a) = 0$ if $a = 0$, and $\text{sign}(a) = 1$ if $a > 0$. It can be shown (Kendall, 1970) that, under the assumption of independent and identically distributed data, expectation $E(S) = 0$ and variance $\text{Var}(S) = \dfrac{n(n-1)/(2n+5)}{18}$. The distribution of S then is asymptotically normal; that is, when n becomes large, the distribution of S can be approximated well by a Gaussian distribution with mean $E(S)$ and variance $\text{Var}(S)$. When we calculate the test statistic[27]

$$\sigma = \frac{S - \text{sign}(S)}{\sqrt{\text{Var}(S)}}, \qquad (4.5)$$

we reject the null-hypothesis "x has no trend" if the test statistic σ derived from observations x is within the 95% percentile range, or any other suitably chosen range, of σ-values which are derived from all sets of n observations that have no trend, suffer no other instationarities, and where the whole set of n observations represents a set of independent samples. The Mann–Kendall test is widely used in geosciences, but there are two problems associated with it. First, the conditions mentioned above are almost never satisfied, because almost all data are indeed serially correlated (i.e., observations are not independent of each other). Second, the conclusion reached by rejecting the null hypothesis is often false, as it represents a change from statistical language to everyday language.

The consequences of disregarding the assumption of serial independence of the data when applying the Mann–Kendall test have been demonstrated by Kulkarni and von Storch (1995). Even if the correlation of two neighboring observations is as low as 10%, the test rejects the null hypothesis of "no trend" much more often than stipulated by the prescribed risk of 5%, or whatever was decided to be acceptable. A solution suggested to overcome this problem is "pre-whitening" of the time series; that is, to first fit an autoregressive process of order 1, and subsequently to subtract this component from the original data. This operation also diminishes the trend that is sought from the data; as a result, a debate is going on—in particular, in hydrology—to discover whether such a procedure is meaningful. In this regard, more recently Hamed (2009) demonstrated the problem for the more severe case when the time series exhibit "long memory".[28] He also discussed improved pre-whitening procedures, all of which have their limitations, however.

The other issue is: What does acceptance of the presence of a trend imply? It means that if we redo the sampling exercise and resample n observations in a parallel

---

[27] $\text{sign}(S)$ was introduced by Kendall (1970) as a continuity correction.
[28] This is a technical term in time series analysis. "Short memory" refers to time series whose auto-corelation function decays relatively fast, with $\rho(\Delta \approx a^{|\Delta|})$, whereas "long memory" time series have a slowly decaying autocorrelation function $\rho(\Delta) \approx |\Delta|^{-\gamma}$ with a suitable γ.

world,[29] then we may expect to see a similar trend in the resampled data. In other words, we make a statement about the *random* experiment in which the n observations are made *again* under similar conditions as for the original sample. This does not imply that such a trend would continue into the future (i.e., after the nth observation), or has prevailed in the past (i.e., before the first sample). Rejection of the null hypothesis refers exclusively to the interpretation of n numbers $x_1, \ldots, x_n$! The often tacitly made conclusion that the trend, once identified as significant, would extend into the future is specious. Statistical significance is not necessarily related to physical significance.

A final word of caution relates to the other assumption behind the Mann–Kendall test; namely, that the time series be stationary. If the time series has a break point, with a sudden change in the mean, for instance, or exhibits cyclostationary behavior, with the mean value undergoing a fixed cycle such as the seasonal cycle, then the time series is not stationary. When a Mann–Kendall test is applied to such data, the null hypothesis is rejected, but the interpretation of an ongoing monotonous change, as implied in the word "trend" of everyday language, is invalid.

4.8 SUMMARY

A number of data quality problems exist that are highly relevant for analyses of long-term changes in marine climate. Issues related to data homogeneity are in many cases severe. The problem arises when the homogeneity of an observational record is compromised by, for example, changes in the accuracy of the instrumental record, in observational practices, or in analysis techniques. Such changes may introduce artificial signals into the observational record that may be interpreted as long-term changes, if they are not properly accounted for. Changing data availability may also cause problems in the interpretation of trends. For example, when geostationary satellite data became available the number of observations over formerly data-sparse oceanic regions was significantly increased. The latter resulted in the detection of more and stronger wind storms which in some cases were attributed to incorrect causes, such as developing anthropogenic climate change. Such data quality problems are not only associated with *in situ* measurements and remotely sensed data but also with derived products such as operational weather analyses.

The usage of proxies is a frequently applied approach to overcome problems associated with data quality. Proxies are data that share the same information as the variable one is interested in, but are less affected by data quality issues. For storminess, an example are indices based on atmospheric pressure gradients, where underlying pressure data are often available for centuries and are usually of good quality, while wind speed measurements are often short and corrupted by inhomogeneities.

---

[29] For instance, the same segment of the annual cycle—say, monthly mean temperatures from March to August—in different years. At least in Europe we would find a significant trend.

Global reanalyses and regional reconstructions represent another option. They can be considered an optimal space–time interpolation of limited observational evidence to derive dynamically and physically consistent pictures of the state of the system using frozen state-of-the-art numerical models and data assimilation techniques. Because of changes in the observational network over time, reanalyses and reconstructions are not completely homogeneous but suffer to some extent from inhomogeneities. The consequences for estimation of long-term changes are mixed, depending on variable, region, and period considered.

The information obtained from proxies and reanalyses complement each other to some extent. While proxies are usually available for very long periods up to a few centuries, they only provide information on and at few scattered locations. Moreover, they require good-quality measurements made over very long periods. Reanalyses and reconstructions are available for much shorter periods (presently up to about 60 years). They provide gridded, physically and dynamically consistent data even for places and variables for which no observations have been made.

In order to derive regional-scale information from global reanalyses or anthropogenic climate change simulations, a number of regionalization techniques have been developed. In dynamical downscaling approaches high-resolution regional models are nested in coarser global or larger scale models in order to enhance resolution at the regional scale. Statistical downscaling may be used in case of limited available computing power, when some observational data on the regional scale or some results from dynamical downscaling are already available. When sufficient computational resources are available, high-resolution global models may represent an alternative.

The question on whether or not such regionalization approaches add value to information that is already available has long been mostly ignored. There have been claims that finer grid spacing (in the case of dynamical downscaling) *per se* adds value and increases performance in comparison with the driving large-scale data. Only recently, a number of studies have emerged that addressed the question in more detail. They have demonstrated that there is indeed added value, but that this added value depends on a number of factors, such as the variable, the scale, or the region considered.

Many problems about future or ongoing long-term changes pose questions about the consequences of such changes and options to deal with them. These types of questions are usually addressed by scenario techniques; that is, by elaborating on consequences in a number of "What … if …?" type of questions that cover many different possible outcomes. In the case of expected anthropogenic climate change the technique is widely applied. Anthropogenic climate change scenarios are built in a series of steps, ranging from greenhouse gas emission scenarios, which are subsequently converted into atmospheric concentrations, to global and regional climate change scenarios that elaborate on the consequences of assumed emissions on large-scale and regional-scale climate. In particular, when regional scales are considered, regionalization techniques are commonly applied. Eventually, marine impact scenarios such as for waves or storm surges may be derived and adaptation strategies and options for coastal protection or shipping may be developed.

The development of adaptation strategies requires some knowledge about ongoing and potential future changes as well as their attribution to causes. The framework applied is generally referred to as detection and attribution, where detection means deciding whether an observed trend is unusual in some sense and attribution means identifying the reasons for change. The latter is of particular interest for the development of response strategies. When the cause of a change is known and to some extent predictable (e.g., the impact of dredging in an estuary on observed storm surge heights), the consequences of different developments can be assessed and appropriate measures can be implemented.

4.9   REFERENCES

Alexandersson, H.; and A. Moberg (1997). Homogenization of Swedish temperature data, Part I: Homogeneity test for linear trends. *Int. J. Climatol.*, **17**, 25–34.

Alexandersson, H.; T. Schmith; K. Iden; and H. Tuomenvirta (1998). Long-term variations of the storm climate over NW Europe. *Global Atmos. Oc. System*, **6**, 97–120.

Alexandersson, H.; H. Tuomenvirta; T. Schmith; and K. Iden (2000). Trends of storms in NW Europe derived from an updated pressure data set. *Climate Res.*, **14**, 71–73.

Auer, I.; R. Boehm; A. Jurkovic; W. Lipa; A. Orlik; R. Potzmann; W. Schoener; M. Ungersboeck; C. Matulla; K. Briffa *et al.* (2005). HISTALP historical instrumental climatological surface time series of the greater alpine region. *Int. J. Climatol.*, **17**, 14–46.

Bärring, L.; and K. Fortuniak (2009). Multi-indices analysis of southern Scandinavian storminess 1780–2005 and links to interdecadal variations in the NW Europe–North Sea region. *Int. J. Climatol.*, **29**, 373–384, doi: 10.1002/joc.1842.

Bärring, L.; and H. von Storch (2004). Scandinavian storminess since about 1800. *Geophys. Res. Lett.*, **31**, L20202, doi: 10.1029/2004GL020441.

Behringer, W. (2005). The global ocean data assimilation system (GODAS) at NCEP. In: *Preprints, 11th Symp. on Integrated Observing and Assimilation Systems for the Atmosphere, Oceans and Land Surface*. American Meteorological Society. Available at *http://ams.confex.com/ams/pdfpapers/119541.pdf*

Bell, M.; J. Martin; and N. Nichols (2004). Assimilation of data into an ocean model with systematic errors near the equator. *Quart. J. Roy. Meteorol. Soc.*, **130**, 853–871.

Bellucci, A.; S. Masina; P. Di Pietro; and A. Navarra (2007). Using temperature–salinity relations in a global ocean implementation of a multivariate dataassimilation scheme. *Mon. Wea. Rev.*, **135**, 3785–3807.

Bengtsson, L.; S. Hagemann; and K. Hodges (2004). Can climate trends be calculated from reanalysis data? *J. Geophys. Res.*, **109**, doi: 10.1029/2004JD004536.

Bhend, J.; and H. von Storch (2007). Consistency of observed winter precipitation trends in northern Europe with regional climate change projections. *Climate Dyn.*, **31**, 17–28, doi: 10.1007/s00382-007-0335-9.

Bhend, J.; and H. von Storch (2009). Is greenhouse gas forcing a plausible explanation for the observed warming in the Baltic Sea catchment area? *Boreal Environ Res.*, **14**, 81–88.

Black, D.; M. Abahazi; R. Thunell; A. Kaplan; E. Tappa; and L. Peterson (2007). An 8-century tropical Atlantic SST record from the Cariaco Basin: Baseline variability, 20th-century warming, and Atlantic hurricane frequency. *Paleooceanography*, **22**, PA4204, doi: 10.1029/2007PA001427.

Bray, D.; C. Hagner; and I. Grossmann (2003). *Grey, Green, Big Blue: Three Regional Development Scenarios Addressing the Future of Schleswig-Holstein*, Technical Report 2003/25. GKSS Research Center, Geesthacht, Germany.

Bromwich, D.; R. Fogt; K. Hodges; and J. Walsh (2007). Tropospheric assessment of ERA-40, NCEP, and JRA-25 global reanalyses in the polar regions. *J. Geophys. Res.*, D11111, doi: 10.1029/2006JD007859.

Caires, S.; and A. Sterl (2005). 100-year return value estimates for ocean wind speed and significant wave height from the ERA-40 data. *J. Climate*, **18**, 1032–1048.

Caires, S.; A. Sterl; and C. Gommenginger (2005). Global ocean mean wave period data: Validation and description. *J. Geophys. Res.*, **110**, C02003, doi: 10.1029/2004JC002631.

Carton, J.; and B. Giese (2008). A reanalysis of ocean climate using simple data assimilation (SODA). *Mon. Wea. Rev.*, **136**, 2999–3017.

Carton, J.; and A. Santorelli (2008). Global decdal upper-ocean heat content as viewed in nine analyses. *J. Climate*, **21**, 6015–6035, doi: 10.1175/2008JCLI2489.1.

Cieślikiewicz, W.; and B. Paplińska-Swerpel (2008). A 44-year hindcast of wind wave fields over the Baltic Sea. *Coastal Eng.*, 894–905, doi: 10.1016/j.coastaleng.2008.02.017.

Compo, G.; J. Whitaker; and P. Sardeshmukh (2006). Feasibility of a 100-year reanalysis using only surface pressure data. *Bull. Amer. Meteorol. Soc.*, **87**, 175–190, doi: 10.1175/BAMS-87-2-175.

Cox, A.; and V. Swail (2001). A global wave hindcast over the period 1958–1997: Validation and climate assessment. *J. Geophys. Res.*, **106**, 2313–2329.

Davey, M. (2005). *Enhanced Ocean Data Assimilation and Climate Prediction*, Framework 5 Project Technical Report. European Commission.

De Kraker, A. (1999). A method to assess the impact of high tides, storms and storm surges as vital elements in climate history: The case of stormy weather and dikes in the northern part of Flanders. *Climatic Change*, **43**, 287–302, doi: 10.1023/A:1005598317787.

Dickinson, R.; R. Errico; F. Giorgi; and G. Bates (1989). A regional climate model for the western United States. *Climatic Change*, **15**, 383–422, doi: 10.1007/BF00240465.

Essen, J.; J. Klussmann; R. Herber; and I. Grevemeyer (1999). Does microseism in Hamburg (Germany) reflect the wave climate in the North Atlantic? *D. Hydrogr. Z.*, **51**, 17–29.

Feser, F. (2006). Enhanced detectability of added value in limited-area model results separated into different spatial scales. *Mon. Wea. Rev.*, **134**, 2180–2190.

Feser, F.; R. Weisse; and H. von Storch (2001). Multi-decadal atmospheric modeling for Europe yields multi-purpose data. *EOS Trans.*, **82**, 305, 310.

Flather, R.; J. Smith; J. Richards; C. Bell; and D. Blackman (1998). Direct estimates of extreme storm surge elevations from a 40-year numerical model simulation and from observations. *Global Atmos. Oc. System*, **6**, 165–176.

García, R.; L. Gimeno; E. Hernández; R. Prieto; and P. Ribera (2000). Reconstructions of North Atlantic atmospheric circulation in the 16th, 17th and 18th centuries from historical sources. *Climate Res.*, **14**, 147–151.

García-Sotillo, M.; A. Ratsimandresy; J. Carretero; A. Bentamy; F. Valero; and F. González-Rouco (2005). A high-resolution 44-year atmospheric hindcast for the Mediterranean Basin: Contribution to the regional improvement of global reanalysis. *Climate Dyn.*, 219–236, doi: 10.1007/s00382-005-0030-7.

Gibson, R.; P. Kålberg; and S. Uppala (1996). The ECMWF re-analysis (ERA) project. *ECMWF Newsl.*, **73**, 7–17.

Giorgi, F. (1990). Simulation of regional climate using a limited area model nested in a general circulation model. *J. Climate*, **3**, 941–963.

Giorgi, F.; B. Hewitson; J. Christensen; M. Hulme; H. von Storch; P. Whetton; R. Jones; L. Mearns; C. Fu; R. Arrit *et al.* (2001). Regional climate information: Evaluation and projections. *IPCC Climate Change 2001: The Scientific Basis.* Cambridge University Press.

Glickman, T. (Ed.) (2000). *Glossary of Meteorology.* American Meteorological Society, Boston, Second Edition.

Gorman, R.; K. Bryan; and A. Laing (2003). Wave hindcast for the New Zealand region: Nearshore validation and coastal wave climate. *N.Z. J. Marine Freshwater Res.*, **37**, 567–588.

Grossmann, I.; K. Woth; and H. von Storch (2006). Localization of global climate change: Storm surge scenarios for Hamburg in 2030 and 2085. *Die Küste*, **71**, 169–182.

Günther, H.; W. Rosenthal; M. Stawarz; J. Carretero; M. Gómez; I. Lozano; O. Serrano; and M. Reistad (1998). The wave climate of the Northeast Atlantic over the period 1955–1994: The WASA wave hindcast. *Global Atmos. Oc. System*, **6**, 121–164.

Hamed, K. (2009). Enhancing the effectiveness of prewhitening in trend analysis of hydrologic data. *J. Hydrol.*, **368**, 143–155, doi: 10.1016/j.jhydrol.2009.01.040.

Hasselmann, K. (1979). On the signal-to-noise problem in atmospheric response studies. In: B. Shaw (Ed.), *Meteorology over the Tropical Oceans.* Royal Meteorological Society, Bracknell, U.K., pp. 251–259.

Hasselmann, K. (1993). Optimal fingerprints for the detection of time dependent climate change. *J. Climate*, **6**, 1957–1972.

Hegerl, H.; H. von Storch; K. Hasselmann; B. Santer; U. Cubasch; and P. Jones (1996). Detecting greenhouse-gas-induced climate change with an optimal fingerprint method. *J. Climate*, **9**, 2281–2306.

Houghton, J.; Y. Ding; D. Griggs; M. Noguer; P. van der Linden; X. Dai; K. Maskell; and C. Johnson (Eds.) (2001). *Climate Change 2001: The Scientific Basis. Contribution of Working Group I to the Third Assessment Report of the Intergovernmental Panel on Climate Change.* Cambridge University Press, Cambridge, U.K. ISBN 0521 01495 6, 881 pp.

IDAG (The International Ad Hoc Detection and Attribution Group) (2005). Detecting and attributing external influences on the climate system: A review of recent advances. *J. Climate*, **18**, 1291–1314.

Ishii, M.; M. Kimoto; K. Sakamoto; and S. Iwasaki (2006). Steric sea level changes estimated from historical ocean subsurface temperature and salinity analyses. *J. Oceanogr.*, **62**, 155–170.

Jedrasik, J.; W. Cieślikiewicz; M. Kowalewski; K. Bradtke; and A. Jankowski (2008). 44-year hindcast of the sea level and circulation in the Baltic Sea. *Coastal Eng.*, 849–860, doi: 10.1016/j.coastaleng.2008.02.026.

Kalnay, E.; M. Kanamitsu; R. Kistler; W. Collins; D. Deaven; L. Gandin; M. Iredell; S. Saha; G. White; J. Woollen *et al.* (1996). The NCEP/NCAR reanalysis project. *Bull. Amer. Meteorol. Soc.*, **77**, 437–471.

Kanamaru, H.; and M. Kanamitsu (2006). Scale selective bias correction in a downscaling of global analysis using a regional model. *Mon. Wea. Rev.*, **135**, 334–350.

Kanamaru, H.; and M. Kanamitsu (2007). 57-year California reanalysis downscaling at 10 km (CaRD10), Part II: Comparison with North American regional reanalysis. *J. Climate*, **20**, 5553–5571.

Kanamitsu, M.; and H. Kanamaru (2007). 57-year California reanalysis downscaling at 10 km (CaRD10), Part I: System detail and validation with observations. *J. Climate*, **20**, 5527–5552.

Kanamitsu, M.; W. Ebisuzaki; J. Woollen; S. Yang; J. Hnilo; M. Fiorino; and G. Potter (2002). NCEP/DOE AMIP-II Reanalysis (R-2). *Bull. Amer. Meteorol. Soc.*, **83**, 1631–1643.

References 161

Karl, T.; R. Quayle; and P. Groisman (1993). Detecting climate variations and change: New challenges for observing and data management systems. *J. Climate*, **6**, 1481–1494.

Kellow, A. (2007). *Science and Public Policy: The Virtuous Corruption of Virtual Environmental Science.* Edward Elgar. ISBN 978-1847204707.

Kendall, M. (1970). *Rank Correlation Methods.* Griffin, London, Fourth Edition, 258 pp.

Kistler, R.; E. Kalnay; W. Collins; S. Saha; G. White; J. Woollen; M. Chelliah; W. Ebisuzaki; M. Kanamitsu; V. Kousky *et al.* (2001). The NCEP/NCAR 50-year reanalysis: Monthly means CD-ROM and documentation. *Bull. Amer. Meteorol. Soc.*, **82**, 247–267.

Köhl, A.; and D. Stammer (2008). Decadal sea level changes in the 50-year GECCO ocean synthesis. *J. Climate*, **21**, 1876–1860.

Kulkarni, A.; and H. von Storch (1995). Monte Carlo experiments on the effect of serial correlation on the Mann–Kendall test of trend. *Meteorol. Z.*, **4**, 82–85.

Landsea, C. (2007). Counting Atlantic tropical cyclones back to 1900. *EOS Trans.*, **88**, 197–208, doi: 10.1029/2007EO180001.

Landsea, C.; C. Anderson; N. Charles; G. Clark; J. Dunion; J. Fernández-Patagás; P. Hungerford; C. Neumann; and M. Zimmer (2004). The Atlantic hurricane database reanalysis project: Documentation for 1851–1910 alterations and additions to the HURDAT data base. In: R. Murnane and K. Liu (Eds.), *Hurricanes and Typhoons: Past, Present and Future.* Columbia University Press, pp. 177–221.

Langenberg, H.; A. Pfizenmayer; H. von Storch; and J. Sündermann (1999). Storm-related sea level variations along the North Sea coast: Natural variability and anthropogenic change. *Continental Shelf Res.*, **19**, 821–842.

Levitus, S.; J. Antonov; and T. Boyer (2005). Warming of the world ocean, 1955–2003. *Geophys. Res. Lett.*, **32**, L02604, doi: 10.1029/2004GL021592.

Luterbacher, J.; E. Xoplaki; D. Dietrich; R. Rickli; J. Jacobeit; C. Beck; D. Gyalistras; C. Schmutz; and H. Wanner (2002). Reconstruction of sea level pressure fields over the eastern North Atlantic and Europe back to 1500. *Climate Dyn.*, **18**, 545–561.

Mann, H. (1945). Nonparametric test against trends. *Econometrica*, **13**, 245–259.

Matulla, C.; and H. von Storch (2009). Changes in eastern Canadian storminess since 1880. *J. Climate*, submitted.

Matulla, C.; W. Schöner; H. Alexandersson; H. von Storch; and X. Wang (2008). European storminess: Late nineteenth century to present. *Climate Dyn.*, **31**, 1125–1130, doi: 10.1007/s00382-007-0333-y.

McGregor, H.; M. Dimma; H. Fischer; and S. Mulitza (2007). Rapid 20th-century increase in coastal upwelling off northwest Africa. *Science*, **315**, 637–639.

Meier, H.; B. Broman; and E. Kjellström (2004). Simulated sea level in past and future climates of the Baltic Sea. *Climate Res.*, **27**, 59–75.

Mesinger, F.; G. DiMego; E. Kalnay; K. Mitchell; P. Shafran; W. Ebisuzaki; D. Jović; J. Woollen; E. Rogers; E. Berbery *et al.* (2006). North American regional reanalysis. *Bull. Amer. Meteorol. Soc.*, **87**, 343–360, doi: 10.1175/BAMS-87-3-343.

Mitchell, J.; D. Karoly; G. Hegerl; F. Zwiers; A. Allen; J. Marengo; V. Baros; M. Berliner; G. Boer; T. Crowley *et al.* (2001). Detection of climate change and attribution of causes. In: *Climate Change 2001: The Scientific Basis. Contribution of Working Group I to the Third Assessment Report of the Intergovernmental Panel on Climate Change.* Cambridge University Press, Cambridge, U.K., pp. 695–738. ISBN 0521 01495 6.

Moberg, A.; and H. Alexandersson (1997). Homogenization of Swedish temperature data, Part II: Homogenized gridded air temperature compared with a subset of global gridded air temperature since 1861. *Int. J. Climatol.*, **17**, 35–54.

Musić, S.; and S. Nicković (2008). 44-year wave hindcast for the eastern Mediterranean. *Coastal Eng.*, 872–880, doi: 10.1016/j.coastaleng.2008.02.024.

Nakicenovic, N.; and R. Swart (Eds.), *Special Report of the Intergovernmental Panel on Climate Change on Emission Scenarios.* Cambridge University Press. Available at *http://www.ipcc.ch/pub/reports/htm.*

Paciorek, C.; J. Risbey; V. Ventura; and R. Rosen (2002). Multiple indices of Northern Hemisphere cyclone activity, winters 1949–99. *J. Climate*, **15**, 1573–1590.

Peterson, E.; and L. Hasse (1987). Did the Beaufort scale or the wind climate change? *J. Phys. Oceanogr.*, **17**, 1071–1074.

Ratsimandresy, A.; M. Sotillo; J. Carretero; E. Alvarez; and H. Hajji (2008). A 44-year high-resolution ocean and atmospheric hindcast for the Mediterranean basin developed within the HIPOCAS project. *Coastal Eng.*, 827–842, doi: 10.1016/j.coastaleng.2008.02.025.

Rybski, D.; A. Bunde; S. Havlin; and H. von Storch (2006). Long-term persistence in climate and the detection problem. *Geophys. Res. Lett.*, **33**, L06718, doi: 10.1029/2005GL025591.

Schmidt, H.; and H. von Storch (1993). German Bight storms analysed. *Nature*, **365**, 791.

Shepherd, J.; and T. Knutson (2007). The current debate on the linkage between global warming and hurricanes. *Geography Compass*, **1**, 1–24, doi: 10.1111/j.1749-8198. 2006.00002.x.

Smits, A.; A. K. Tank; and G. Können (2005). Trends in storminess over the Netherlands, 1962–2002. *Int. J. Climatol.*, **25**, 1331–1344, doi: 10.1002/joc.1195.

Solomon, S.; D. Qin; M. Manning; Z. Chen; M. Marquis; K. Averyt; M. Tignor; and H. Miller (Eds.) (2007). *Climate Change 2007: The Physical Science Basis. Contribution of Working Group I to the Fourth Assessment Report of the Intergovernmental Panel on Climate Change.* Cambridge University Press, Cambridge, U.K., 996 pp. ISBN 978-0-521-88009-1.

Sterl, A.; and S. Caires (2005). Climatology, variability and extrema of ocean waves: The web-based KNMI/ERA-40 wave atlas. *Int. J. Climatol.*, **25**, 963–977, doi: 10.1002/joc.1175.

Sterl, A.; G. Komen; and P. Cotton (1998). Fifteen years of global wave hindcasts using winds from the European Centre for Medium-range Weather Forecasts reanalysis: Validating the reanalyzed winds and assessing wave climate. *J. Geophys. Res.*, **103**, 5477–5492.

Stott, P. (2003). Attribution of regional-scale temperature changes to anthropogenic and natural causes. *Geophys. Res. Lett.*, **30**, doi: 10.1029/2003GL017324.

Sun, C.; M. Rienecker; A. Rosati; M. Harrison; A. Wittenberg; C. Keppenne; J. Jacob; and R. Kovach (2007). Comparison and sensitivity of ODASI ocean analysis in the tropical Pacific. *Mon. Wea. Rev.*, **135**, 2242–2264.

Tol, R. (2006). Exchange rates and climate change: An application of FUND. *Climate Change*, **75**, 59–80.

Tol, R. (2007). Economic scenarios for global change. In: H. von Storch, R. Tol, and G. Flöser (Eds.), *Environmental Crisis: Science and Policy.* Springer-Verlag, pp. 17–36. ISBN 978-3-40-75895-2.

Trenberth, K.; P. Jones; P. Ambenje; R. Bojariu; D. Easterling; A. K. Tank; D. Parker; F. Rahimzadeh; J. Renwick; M. Rusticucci *et al.* (2007). *Climate Change 2007: The Physical Science Basis. Contribution of Working Group I to the Fourth Assessment Report of the Intergovernmental Panel on Climate Change.* Cambridge University Press, Cambridge, U.K. ISBN 978-0-521-99009-1.

Uppala, S.; P. Kallberg; A. Simmons; U. Andrae; V. da Costa Bechtold; M. Fiorino; J. Gibson; J. Haseler; A. Hernández; G. Kelly *et al.* (2005). The ERA-40 re-analysis. *Quart. J. Roy. Meteorol. Soc.*, **131**, 2961–3012.

von Storch, H. (2007). Climate change scenarios: Purpose and construction. In: H. von Storch, R. Tol, and G. Flöser (Eds.), *Environmental Crisis: Science and Policy*. Springer-Verlag, pp. 5–16. ISBN 978-3-540-75895-2.

von Storch, H.; and H. Reichardt (1997). A scenario of storm surge statistics for the German Bight at the expected time of doubled atmospheic carbon dioxide concentration. *J. Climate*, **10**, 2653–2662.

von Storch, H.; and R. Weisse (2008). Regional storm climate and related marine hazards in the northeast Atlantic. In: H. Diaz and R. J. Murnane (Eds.), *Climate Extremes and Society*. Cambridge University Press, pp. 54–73. ISBN 978-0-521-87028-3.

von Storch, H.; and F. Zwiers (1999). *Statistical Analysis in Climate Research*. Cambridge University Press, New York, 494 pp.

von Storch, H.; E. Zorita; and U. Cubasch (1991). *Downscaling of Global Climate Change Estimates to Regional Scales: An Application to Iberian Rainfall in Wintertime*, MPI Report 64. Max-Planck-Institut für Meteorologie, Hamburg, Germany.

von Storch, H.; H. Langenberg; and F. Feser (2000). A spectral nudging technique for dynamical downscaling purposes. *Mon. Wea. Rev.*, **128**, 3664–3673.

Wang, X.; V. Swail; F. Zwiers; X. Zhang; and Y. Feng (2009). Detection of external influence on trends of atmospheric storminess and northern ocean wave heights. *Climate Dyn.*, **32**, 189–203, doi: 10.1007/s00382-008-0442-2.

WASA-Group (1998). Changing waves and storms in the northeast Atlantic? *Bull. Amer. Meteorol. Soc.*, **79**, 741–760.

Weisse, R.; and H. Günther (2007). Wave climate and long-term changes for the southern North Sea obtained from a high-resolution hindcast 1958–2002. *Ocean Dynamics*, **57**, 161–172, doi: 10.1007/s10236-006-0094-x.

Weisse, R.; and A. Pluess (2006). Storm-related sea level variations along the North Sea coast as simulated by a high-resolution model 1958–2002. *Ocean Dynamics*, **56**, 16–25, doi: 10.1007/s10236-005-0037-y, on line 2005.

Weisse, R.; H. von Storch; U. Callies; A. Chrastansky; F. Feser; I. Grabemann; H. Günther; A. Pluess; T. Stoye; J. Tellkamp *et al.* (2009). Regional meteorological-marine reanalyses and climate change projections: Results for Northern Europe and potentials for coastal and offshore applications. *Bull. Amer. Meteorol. Soc.*, **90**, 849–860, doi: 10.1175/2008BAMS2713.1.

Winterfeldt, J. (2008). *Comparison of Measured and Simulated Wind Speed Data in the North Atlantic*, GKSS Report 2008/2. GKSS Forschungszentrum, Geesthacht, Germany.

Winterfeldt, J.; and R. Weisse (2009). Using QuickSCAT in the added value assessment of dynamically downscaled wind speed. *Int. J. Climatol.*, submitted.

Woodworth, P.; and D. Blackman (2002). Changes in extreme high waters at Liverpool since 1768. *Int. J. Climatol.*, **22**, 697–714.

Zorita, E.; and H. von Storch (1999). The analog method as a simple statistical downscaling technique: Comparison with more complicated methods. *J. Cimate*, **12**, 2474–2489.

Zwiers, F.; and H. von Storch (2004). On the role of statistics in climate research. *Int. J. Climatol.*, **24**, 665–680.

5

Past and future changes in wind, wave, and storm surge climates

5.1 INTRODUCTION

In this chapter we review present knowledge about past and potential future changes and variability of marine weather phenomena. The review provides a snapshot of knowledge as of early 2009 when this chapter was completed. Conclusions, numbers, and interpretations may change when new observational evidence becomes available or when some of the identified shortcomings have been addressed.

In the following we often emphasize that different studies or models have provided *consistent* results, meaning that there is a *consensus* emerging. Scientific consensus is not by itself a scientific argument, although it is often used that way by many scientists. Scientific consensus merely represents a collective judgment or a position of a group of scientists. It is usually achieved through replication of results, publications, communications at conferences, etc. Scientific consensus may change over time when new insights and results emerge. For example, the theory of continental drift proposed by Alfred Wegener was rejected by most geologists at the time it was presented. Nowadays, it is widely accepted and part of scientific consensus. The results presented in this chapter should be interpreted that way. They reflect our present knowledge and theories, which may be supported or partly refused when new evidence emerges.

When potential future changes are addressed, uncertainties are often assessed. These are not uncertainties in the classical sense of error statistics. When anthropogenic climate change is considered, uncertainties may arise from several sources. First, there is uncertainty related to future social and economic developments. This type of uncertainty is usually addressed by constructing emission scenarios based on anticipated socio-economic developments (see Section 4.6). In this way a considerable range of possible future developments may be captured. There is, however, no guarantee that the entire space of possible developments is fully or evenly sampled. In other words, we do not know whether *all* possible developments are captured, nor whether scenarios provide an independent and identically distributed sample from the entire sample space. The latter is problematic in cases where probability statements and functions are derived, such as in the latest IPCC report (Solomon *et al.*, 2007).

A second type of uncertainty arises from the use of models in our assessment of changing future climate conditions. Models are a manifestation of our present knowledge and understanding of the climate system. Their solutions may deviate from each other because of inherent natural (model) variability, or because some processes are treated differently by different models. Provided that all climate models represent the state of the art, differences in their results thus reflect not only internal variability but also deficiencies in our knowledge. To address this type of uncertainty multi-model ensembles are usually considered. The range of solutions offered by these ensembles is then interpreted as our uncertainty in knowledge. Again, it cannot be proven whether the suite of models applied provides an independent and identically distributed sample from the entire sample space. Again, care is required when probability statements are derived. Generally, the uncertainties estimated in that way represent *lower bounds*; that is, larger errors cannot be excluded.

In the following we briefly review present knowledge on past and future changes in marine weather phenomena. We begin with mid-latitude storms and storm tracks (Section 5.2). Subsequently, changes in tropical cyclone statistics are addressed (Section 5.3). In Section 5.4 results for ocean wave statistics are described. Tides, storm surges, and mean sea level are discussed in Section 5.5.

5.2 MID-LATITUDE CYCLONES AND STORM TRACKS

5.2.1 Past changes and variability

Extra-tropical or mid-latitude cyclones are the dominant weather phenomenon in mid-latitudes (approx. 30–60°N) during the cold season. They form on strong meridional temperature gradients and transport energy towards the poles, thereby reducing the temperature gradient that produced them (Section 2.2). Extra-tropical cyclones tend to occur within restricted regions usually referred to as storm tracks. In the Northern Hemisphere, two major storm tracks are found over the North Pacific and the North Atlantic oceans, while in the Southern Hemisphere a zonally more symmetric storm track spanning the entire latitude belt is found. Secondary storm tracks are found at some places, such as over the Mediterranean Sea.

Studies on changes in extra-tropical storm activity usually focus on changes in the number, intensity, or tracks of mid-latitude cyclones. Most studies are based on reanalysis data limiting the time period to the second half of the 20th century and credibility is restricted by the limitations inherent in the reanalyses (see Section 4.4). This makes detection of any (anthropogenic) trend difficult; in particular, as variability is high in relation to the observed trend and dependent on and associated with large-scale atmospheric circulation modes. In addition, some studies using more homogeneous proxy data (Section 4.3) are available, extending the time period that can be considered and allowing assessment of the credibility of signals estimated from reanalysis data. Such proxies are, however, available only for a limited number of places on the globe, mainly in Europe, providing a spatially confined picture.

Seasonally, storm activity peaks in late fall within the North Pacific and in January within the North Atlantic storm track (Nakamura, 1992). While there is still ongoing discussion about the mechanisms of this difference, Nakamura (1992) showed that baroclinic wave activity is suppressed when jet stream wind speeds exceed about $45\,\mathrm{m\,s^{-1}}$, a situation that usually occurs in winter over the North Pacific, but which is rarely found over the Atlantic Basin. There is also seasonal variation in the location of both storm tracks with more poleward/equatorward locations occurring during the summer/winter months.

A large number of studies deal with inter-annual and decadal variability in the North Pacific and North Atlantic storm tracks. Chang and Fu (2002) computed the leading EOF[1] modes of storm track variability based on 51 NCEP/NCAR reanalysis (Section 4.4) winters (1948/1949–1998/1999). They found that the dominant mode of variability, accounting for about 29% of variance, represents a simultaneous change in amplitude (intensity) of both major Northern Hemisphere storm tracks (Figure 5.1). The corresponding principal component time series, describing the time evolution of this pattern, was found to show pronounced inter-decadal variability with mean storm track intensities being nearly 30% stronger during the late 1990s than the late 1960s/early 1970s (Figure 5.1). This corresponds to the results of other studies, likewise based on reanalysis data. For a review of such studies see, for example, Chang *et al.* (2002).

Harnik and Chang (2003) analyzed storm track variations as seen in reanalysis data and compared them with radiosonde observations. They concluded that inter-annual to decadal storm track variability in the Northern Hemisphere is reasonably reproduced by reanalysis data, while the radiosonde data suggest an exaggeration of long-term trends derived from reanalysis. Harnik and Chang (2003) suggested that this is due to an overall decrease in reanalysis biases with time. A similar conclusion was reached by Chang (2007) who assessed and compared trends in Northern Hemisphere winter storm track activity based on reanalyses and surface ship observations. In particular, he found that, even after corrections were made to the ship data to account for secular changes in observational error statistics, ship-based trends in the North Pacific are much smaller than those found in the reanalyses, while over the Atlantic corrected ship-based trends were consistent with those obtained from reanalyses.

A number of studies relates observed storm track and extra-tropical cyclone variability to different large-scale modes of variability such as the El Niño–Southern Oscillation, the North Atlantic Oscillation, or the Arctic Oscillation. Most authors report more or less robust relations such as an equatorward and downstream shift of North Pacific storm track activity (e.g., Chang *et al.*, 2002), or enhanced U.S. East Coast storm activity during El Niño years (Eichler and Higgins, 2006), or a poleward shift of North Atlantic cyclone activity during positive phases of the North Atlantic

---

[1] Empirical orthogonal functions, see von Storch and Zwiers (1999) for details.

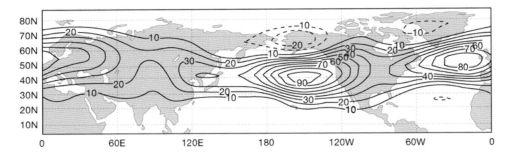

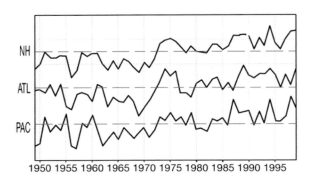

Figure 5.1. Spatial pattern of the leading EOF mode of storm track activity derived from 300 hPa bandpass-filtered velocity variance in $m^2 s^{-2}$ (top) and corresponding principal component time series (dimensionless) for the Northern Hemisphere (NH), the North Atlantic (ATL), and the North Pacific (PAC) (bottom). Redrawn after Chang and Fu (2002). Reproduced/modified by permission of American Meteorological Society.

Oscillation or the Arctic Oscillation (e.g., Serreze et al., 1997). Chang and Fu (2002) showed, however, that even after contributions from large-scale modes of variability were removed from storm activity data, their residuals still exhibit significant interdecadal variations. This indicates that, while some fraction of the observed variability in extra-tropical storm activity may be explained by corresponding variations in large-scale atmospheric or coupled modes, there remains a considerable fraction unaccounted for, being the result of other mechanisms.

A number of recent studies report a noticeable shift in Northern Hemisphere storm tracks, increased storm activity, and a decrease in the number of extra-tropical cyclones during the second half of the 20th century. Figure 5.2 (see color section) shows an example from McCabe et al. (2001) who investigated trends in Northern Hemisphere surface cyclone frequency and intensity 1959–1997 based on reanalysis data. Obviously, cyclone frequency decreased between 30°N and 60°N and increased poleward of 60°N corresponding to a poleward shift of storm track locations. At the same time, cyclone intensity increased in the entire Northern Hemisphere north of 30°N. Corresponding results were obtained from individual analyses of North Atlantic (Geng and Sugi, 2001) storm activity and North Pacific (Graham and Diaz, 2001) storm activity. The results are further corroborated by Paciorek et al. (2002) who showed that reanalysis wind speeds increased between 20°N and 70°N over the reanalysis period over both the North Pacific Basin and the North Atlantic Basin.

Sec. 5.2] **Mid-latitude cyclones and storm tracks** 169

However, extra-tropical cyclone counts showed few or decreasing trends (Paciorek *et al.*, 2002).

Results from reanalyses are somewhat limited because of the relatively short time period for which such data are available and because of potentially existing time-dependent biases which make interpretation of long-term changes difficult. It is thus necessary to confront these results with those obtained from the analysis of long proxy records, which are available at some places. The longer period for which such proxies are available makes it possible to put statements about reanalysis trends into a longer term perspective as well.

A method for deriving a homogeneous proxy for storm activity in a region was proposed by Schmidt and von Storch (1993). They used a set of homogeneous station pressure data to derive upper geostrophic wind speed percentiles which served as a proxy for storminess in the German Bight (southern North Sea). Analyzing this proxy, they found noticeable inter-annual and decadal fluctuations but no long-term trend over the period 1876–1990 (Figure 4.8). The method was applied more systematically by Alexandersson *et al.* (1998, 2000) for a large number of stations in northwest Europe. The most recent update of this analysis was presented in the *Fourth Assessment Report of the Intergovernmental Panel on Climate Change* (IPCC) (Trenberth *et al.*, 2007). It shows that storm activity in the area was relatively high at the end of the 19th century. It declined until about 1960, subsequently a strong increase was observed. Storm activity again peaked in the late 1990s returning to more moderate and calm conditions in the most recent years (Figure 5.3, see color section). The behavior after 1960 is broadly consistent with that obtained from reanalyses (e.g., Chang and Fu, 2002; Chang, 2007). In a similar way Matulla *et al.* (2008) analyzed upper geostrophic wind speed percentiles from the late 19th century up to the present for northwestern, northern, and central Europe. They found a broad agreement between changes in the different areas, again with high levels of storm activity at the beginning of the 20th century and around 1990, and low levels in the early 21st century and around 1960. Matulla *et al.* (2008) described the behavior as most pronounced in northwestern Europe and less pronounced in central Europe. Two of the longest homogeneous pressure records in Europe are available for Stockholm and Lund in Sweden. Bärring and von Storch (2004) analyzed storm activity based on a number of different indices derived from these data sets. They found that there is pronounced inter-annual and decadal variability but no discernible long-term trend stronger than the inherent long-term variability. In other words, proxy-based storm indices for Europe are broadly consistent with results obtained from reanalyses over reanalysis periods. Analysis of longer time periods for which these proxies are available suggests that trends in storm activity inferred from reanalysis over the past 50 years or so seem to lie very well within the range of observed natural variability.

Long proxy records for storm activity and their analysis for areas other than Europe are less frequent. Matulla and von Storch (2009) analyzed variations in storm activity in eastern Canada based on historical data from about 1880 onwards. For Arctic stations, they found a behavior broadly consistent with that in northwest Europe; namely, a peak around 1920, a decrease in the 1960s, and another increase that comes to an end at the end of the 1990s. The most recent years in the analysis of Matulla and von Storch (2009) are characterized again by calmer conditions. Canadian stations south of the Arctic such as Toronto seem to feature an opposite behavior with observed long-term changes well within the range of observed historical observations. Combining three different data sets for storm frequency and intensity for the Great Lakes area Angel and Isard (1998) reached somewhat different conclusions. They describe an unbroken increasing trend since about 1900 in the number of strong cyclones defined by surface core pressures of less than 992 hPa. According to their analysis, the number of strong cyclones in the area almost doubled during the 20th century. While the results of Matulla and von Storch (2009) and Angel and Isard (1998) are somewhat contradictory, both are subject to considerable uncertainty. Matulla and von Storch (2009) reported on several inhomogeneities in their data, which they had to correct for before deducing the character of changes in storm activity. Angel and Isard (1998) combined three different data sets, and reported that the number of stations used for the detection of pressure minima in the region dramatically increased from 38 in 1901 to 262 in 1980. Unfortunately, no explicit attempt was made to quantify the consequences of these changes in the detection of strong cyclones in the area, making the described trend somewhat suspicious.

There are a number of other proxies related to storm activity. Bromirski *et al.* (2003) analyzed hourly tide gauge data from San Francisco 1858–2000. Focusing on non-tidal residuals, they found that these did not change substantially, neither within the entire period 1858–2000 nor for typical reanalysis periods (1951–2000). Extreme winter non-tidal residuals, however, show an increase over the reanalysis period consistent with results obtained directly from reanalyses (Chang and Fu, 2002). Similar results were obtained (e.g., by Woodworth and Blackman, 2002) from an analysis of tide gauge data in Liverpool since 1768. For a more detailed discussion of these data we refer the reader to Section 5.5.2. Summarizing, the results from many different analyses presently suggest that storm activity in the Northern Hemisphere was not much different in the second half of the 20th century than before.

For the Southern Hemisphere, changes in extra-tropical storm activity are less well documented. Some studies report a decrease in the number of mid-latitude cyclones over typical reanalysis periods (e.g., Simmonds and Keay, 2000). However, for the Southern Hemisphere changes and variability are usually derived from reanalysis data which, in the Southern Hemisphere, have a high degree of uncertainty. Here changes in the design of the observational network—in particular, the advent of satellites—are very likely to introduce time-dependent biases in the reanalyses that may hamper the interpretation of climate variability and trends obtained from such data (e.g., Bengtsson *et al.*, 2004) (see also Section 4.4). Similarly, changes based on near-surface wind speed observations analyzed by some authors are subject to high uncertainties too. Over the oceans such data are mainly from voluntary observing ships and have time-dependent biases (Gulev *et al.*, 2007) such as changes from visual observations to measurements, changes in anemometer heights and types, changes in the Beaufort Scale, growing ship size, etc.

5.2.2 Future changes

The range of studies existing so far has presented a rather mixed picture regarding extra-tropical cyclone changes related to anthropogenic climate change. Reviewing the existing literature, a number of different competing processes and arguments are discussed, the most common of which are

1. *Changing meridional temperature gradients.* Extra-tropical cyclones form and grow via baroclinic instability (Holton, 1992) receiving their kinetic energy from the conversion of available potential energy (Peixoto and Oort, 1992). Available potential energy is proportional to variance of the temperature field (Bengtsson *et al.*, 2009)—that is, the strength of meridional temperature gradients. In the lower troposphere, meridional temperature gradients are generally expected to decrease due to stronger anthropogenic warming in polar regions. In the upper troposphere, they are expected to increase as a result of strong warming in upper-tropospheric levels at low latitudes (e.g., Solomon *et al.*, 2007). The discussion concerns which of the two opposite developments will have a stronger influence on extra-tropical cyclone statistics. Held (1993) stated that dominant wintertime mid-latitude cyclones are coherent throughout the troposphere and concluded that it is unclear whether they will respond to increased upper-tropospheric or decreased lower-tropospheric meridional temperature gradient. Bengtsson *et al.* (2006) recalled that increased upper-tropospheric and decreased lower-tropospheric temperature gradients also implies vertically more stable stratification at lower latitudes and less stable stratification at higher latitudes, which may be responsible for the poleward shift of extra-tropical storm activity and hence the poleward movement of average storm track locations.
2. *Increasing amount of water vapor.* Anthropogenic warming of the atmosphere will be accompanied by an increase in the amount of water vapor broadly following the Clausius–Clapeyron relation (Bengtsson *et al.*, 2006). As a consequence, the atmosphere may become more efficient at carrying out meridional heat transport, requiring less intense vortices to maintain the balance (Bengtsson *et al.*, 2006). Other authors emphasize that the enhanced release of latent heat may contribute to stronger development of extra-tropical cyclones. Chang *et al.* (2002) analyzed storm track dynamics by investigating local transient eddy energy budgets and found that condensational heating is adding about 20% to the baroclinic generation rate over storm track entrance regions. Some authors consider the effect to be minor (e.g., Bengtsson *et al.*, 2009).
3. *Changes in sea surface temperatures.* Changes in sea surface temperatures have been suggested by several authors to enhance extra-tropical cyclone development. Other authors consider it less likely as warmer SSTs act as a sink for available potential energy (Bengtsson *et al.*, 2006). They may, however, play a role in regional modifications of storm track locations (Bengtsson *et al.*, 2006).

It is obvious that these processes are partly competing. Their relative contributions to observed or expected extra-tropical cyclone variability and change

172   **Past and future changes in wind, wave, and storm surge climates**   [Ch. 5

are sometimes disputed in the scientific literature. In experiments with numerical models, the outcome may depend on details in the balance of different processes which may be different in different models. The latter may explain the large range of results obtained regarding extra-tropical cyclone variability and change and illustrates uncertainties related to our incomplete knowledge and understanding of the system. However, some consistent points seem to have emerged more recently:[2]

1. *The poleward shift of extra-tropical storm tracks.* The *Fourth Assessment Report of the Intergovernmental Panel on Climate Change* (IPCC) notes that a tendency for a poleward shift in extra-tropical cyclone activity of several degrees latitude in both hemispheres appears to be a consistent result of many studies that have emerged more recently (Meehl *et al.*, 2007). The shift is associated with increased storm activity at higher latitudes, and reduced storm activity at mid-latitudes. Possible explanations have been put forward (e.g., by Bengtsson *et al.*, 2006 or Yin, 2005) and are related to differential changes of meridional temperature gradient with height and resulting regional differences in changes in vertical stability (see above). Generally, the poleward shift in extra-tropical storm activity appears to be more pronounced in the Southern Hemisphere (Bengtsson *et al.*, 2006). Regionally, large differences may occur as secondary changes in storm track location, and activity may be associated with regional SST changes (e.g., Yin, 2005; Bengtsson *et al.*, 2006).

2. *Fewer extra-tropical cyclones.* While the most recent IPCC report emphasizes the poleward shift of extra-tropical storm track location as a more consistent result (Meehl *et al.*, 2007), earlier reports (Houghton *et al.*, 2001) more strongly noted the possibility that the overall number of extra-tropical cyclones may decrease (e.g., Knippertz *et al.*, 2000; Geng and Sugi, 2003) or show little change (e.g., Kharin and Zwiers, 2005; Watterson, 2005) in response to anthropogenic global warming.

3. *More intense extra-tropical storms.* Similarly, earlier IPCC reports noted the tendency towards more intense mid-latitude cyclones in response to anthropogenic warming (Meehl *et al.*, 2007). Studies indicating that strong extra-tropical storms may become more frequent comprise Pinto *et al.* (2007), Fischer-Bruns *et al.* (2005), Bengtsson *et al.* (2006), or Caires *et al.* (2006). The changes are consistent with the reasoning provided, for example, in Bengtsson *et al.* (2009), although Bengtsson *et al.* (2006) and Sinclair and Watterson (1999) noted that the result may be an artifact of the analysis methods in some studies (e.g., Lambert and Fyfe, 2006) and may partially reflect only changes in the atmospheric large-scale background pressure field.

Summarizing, the most recent IPCC report emphasized that a poleward shift of storm track positions has emerged recently as a more consistent result. Regionally, large differences may occur associated with changes in the position and variability of

---

[2] Note that the degree of consistency is not *per se* evidence that a particular result or outcome is true or false (see Section 5.1).

tracks. Regional details of storm track changes are not well projected, with some authors suggesting that larger regional changes are within the range of observed natural variability or, alternatively, may be the result of local SST changes (e.g., Bengtsson *et al.*, 2009).

A number of studies showed observed signals consistent with the expected patterns of anthropogenic change (e.g., McCabe *et al.*, 2001; Paciorek *et al.*, 2002) (see also Section 5.2.1). Wang *et al.* (2009) presented a detection analysis for trends in geostrophic wind speed indices, mean sea level pressures, and significant ocean wave heights. The analysis was presented for the second half of the 20th century comprising typical reanalysis periods. Wang *et al.* (2009) found that the "trend patterns contain a detectable response to anthropogenic and natural[3] forcing combined." However, they also found the results to be partly conflicting depending on the data sets used. In addition, only the second half of the last century was considered. Other studies based on the analysis of longer proxy records such as those by the WASA-Group (1998) and several others since then (e.g., Bärring and von Storch, 2004) concluded that the observed changes in northern European storm climate are not inconsistent with observed natural variability. Other studies such as that by Fischer-Bruns *et al.* (2005) concluded that solar and volcanic activity only have negligible influence on variations in extra-tropical storm activity. As such, the question on whether or not a noticeable anthropogenic signal can already be detected in present records of observed extra-tropical cyclone activity still remains an open issue.

5.3 TROPICAL CYCLONES

5.3.1 Past changes and variability

The factors determining and affecting tropical cyclone formation are reviewed in Section 2.3. Changes in any of these factors eventually will have an effect on tropical cyclone statistics such as their average number, duration, intensity, or track. The processes are partly competing and their net effect strongly depends on the balance of individual contributions.

A large number of different measures for describing tropical cyclone statistics exist and are used in various studies. The measures all focus on different aspects of tropical cyclone variability which makes the results difficult to compare. For example, early studies often used the number of moderate or intense tropical cyclones detected over a year or a season. The classification is usually based on near-surface *sustained* wind speeds according to the Saphir–Simpson scale. The way sustained wind speeds are determined, however, varies regionally and in time. Recent studies often use large-scale indices such as the genesis potential (Emanuel and Nolan, 2004), the power dissipation index (Emanuel, 2007), or the accumulated cyclone energy index (Bell *et al.*, 2000) to characterize variations in basin-wide or global tropical cyclone statistics. Not only the metrics of these indices, but also the data basis for their

---

[3] Solar and volcanic.

computation varies and is, sometimes seriously, limited by data issues such as changes in the density of the observational network, or cyclone analysis and detection methods (see Section 4.2.2 for details). The use of different metrics and the limitations inherent in the observational data base may account for the range of different results found in the scientific literature. Quantification of past tropical cyclone variability and change still remains a debated issue (e.g., Emanuel, 2007; Landsea, 2005; Chan, 2006).

In the *Fourth Assessment Report of the Intergovernmental Panel on Climate Change* (IPCC) the accumulated cyclone energy index (ACE) is discussed to some extent (Trenberth *et al.*, 2007). The index basically reflects wind energy and is proportional to the square of wind speed. Globally, ACE can be computed only for very short periods when more or less reliable data are available for all ocean basins that show tropical cyclone activity. Klotzbach (2006) provided a global ACE figure for 1986–2005 and described an increasing trend over this 20-year period for the North Atlantic and a decreasing trend for the eastern North Pacific basin. For all other basins, minor changes were reported and, globally, net tropical cyclone activity showed no significant change over the period 1986–2005. The Fourth IPCC Report emphasized that the period analyzed by Klotzbach (2006) was rather short and that no conclusions about longer time scales should be drawn from this study (Trenberth *et al.*, 2007).

For the Northern Hemisphere, an update of the analysis of Klotzbach (2006) was provided by Maue (2009) who analyzed basin-wide and Northern Hemisphere figures of ACE for the period 1981–2007. The results of this study are illustrated in Figure 5.4. The figure indicates relatively constant hemispheric ACE values over the analysis period with large inter-annual variability superimposed and observed during the past three decades. In 2007 and 2008[4] relatively low ACE values were found for the Northern Hemisphere. On a basin scale, tropical cyclone activity was decreasing over the eastern North Pacific and increasing over the North Atlantic with both trends largely compensating. Combined, the activity over both basins has been relatively constant on average, providing about 30%–60% of Northern Hemisphere ACE (Figure 5.4). In total, Northern Hemisphere ACE is dominated by activity in the western Pacific and, as a consequence, it is strongly correlated with Pacific sea surface temperatures (Maue, 2009).

For the western North Pacific, the eastern North Pacific, and the North Atlantic still longer ACE time series were provided by the Fourth IPCC Assessment Report (Trenberth *et al.*, 2007). The time series start around 1950 and were updated through early 2006. Although Trenberth *et al.* (2007) emphasized that reliability improves over time and trends may contain unquantified uncertainties, ACE time series for all three areas do not reveal noticeable long-term changes other than strong inter-annual and decadal fluctuations that are opposing over the eastern North Pacific and the North Atlantic. Trenberth *et al.* (2007) noted a clear El Niño connection in most regions with strong negative correlations between the Pacific and Atlantic regions, such that during El Niño events tropical cyclone activity is typically reduced over the

---

[4] Results of preliminary analysis in Maue (2009) not shown in Figure 5.4.

Sec. 5.3] Tropical cyclones 175

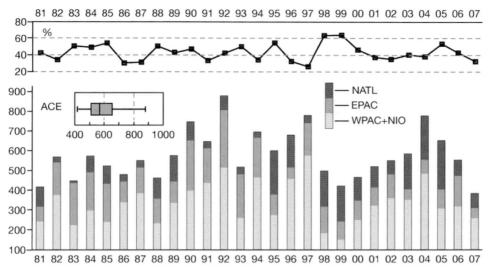

Figure 5.4. Northern Hemisphere and individual basin accumulated tropical cyclone energy (ACE) 1981–2007. NATL denotes the North Atlantic, EPAC the eastern Pacific, WPAC the western Pacific, and NIO the North Indian Ocean. The ratio between EPAC and NATL ACE is given by the time series at the top. The inset box-and-whisker plot shows 1981–2007 global ACE distribution with the median and the upper and lower quartiles. Redrawn from Maue (2009). Published/copyright (2009) American Geophysical Union. Reproduced/modified by permission of American Geophysical Union.

Atlantic, the far West Pacific, and Australian regions while it is increased over the central North and South Pacific, especially over the western North Pacific typhoon region (Gray, 1968).

Analysis of other indices may provide somewhat different pictures. For example, Emanuel (2005) developed a power dissipation index (PDI) for tropical cyclones that basically reflects the cube of maximum sustained wind speeds at 6-hour intervals. Analyzing North Atlantic and western North Pacific PDI since about 1950 Emanuel (2007) concluded that there has been a substantial increase in tropical cyclone activity since about 1970 and that low-frequency North Atlantic PDI variations are highly correlated with low-frequency variations of tropical Atlantic SST. Over the eastern North Pacific decreases in PDI have been reported by other authors. Trenberth *et al.* (2007) noted that, because PDI is proportional to the cube of the wind speed, it may be very sensitive to data issues.

Results of such analyses have been challenged by several studies such as those by Landsea (2005) or Chan (2006), who questioned the quality of the data and suggested alternatives like changes in the observing network, changing data availability, or modification of analysis techniques to be partly responsible for some of the observed changes. A more detailed discussion can be found in Section 4.2.2 or in the original papers of, for example, Landsea *et al.* (2004), Landsea (2005), Chan (2006), or Landsea (2007).

176 **Past and future changes in wind, wave, and storm surge climates** [Ch. 5

For most basins reliable records for tropical cyclone activity are relatively short, making it difficult to determine or assess any ongoing long-term change and variability. Studies that try to detect anthropogenic influences in observed records therefore mainly exist for North Atlantic hurricanes for which the longest record exists. It begins in 1851, is fairly reliable after about 1950 when aircraft reconnaissance emerged (Trenberth *et al.*, 2007), and most reliable after the early 1970s (Landsea, 2005). The North Atlantic hurricane record shows strong multi-decadal fluctuations with a fairly active period around 1930–1960 and a less active period during the 1970s and 1980s (Trenberth *et al.*, 2007). Beginning in the mid-1990s, again a relatively active period emerged and increased North Atlantic hurricane activity correlates well with increased sea surface temperatures in the North Atlantic (Emanuel, 2007). The causes of these changes are currently debated with conflicting results. Goldenberg *et al.* (2001) argued that the time behavior of North Atlantic hurricane activity is mainly oscillatory, being modulated by the Atlantic Multi-decadal Oscillation (AMO), a near-global scale mode of observed multi-decadal climate variability with alternating warm and cool phases of sea surface temperatures that may last for 20–40 years and which comprise large parts of the Northern Hemisphere (Knight *et al.*, 2006). Other authors attribute the changes in North Atlantic sea surface temperatures to variations in the radiative forcing caused by varying solar activity, anthropogenic greenhouse gas, and anthropogenic and volcanic aerosol concentrations (Mann and Emanuel, 2006). The observed increase in North Atlantic hurricane activity after about 1995 may thus represent a combination of the response to sea surface warming as a consequences of anthropogenic climate change and a favorable phase of the AMO (Trenberth *et al.*, 2007). There is, however, still substantial ongoing discussion about the mechanisms of long-term tropical cyclone variability and change as well as about their relative contributions. Quantification of these contributions strongly determines the possible outcome of the discussion; namely, whether we will experience a relatively small change that can hardly be detected for decades into the future or a relatively large change that is noticeable already today.

5.3.2 Future changes

Tropical cyclone statistics may change in a number of different ways in response to greenhouse gas–induced warming. This comprises changes in the frequency, intensity, size, duration, or climatological tracks of cyclones. When potential future changes in tropical cyclone statistics are considered, significant uncertainties arise from the choice of underlying emission scenario, climate (model) sensitivity to these scenarios, and the limited capacity of the present generation of climate models to adequately simulate tropical cyclones; in particular, the most intense ones. Confidence in any future projection of tropical cyclone statistics will thus largely depend on confidence in the projections of local thermodynamic changes (e.g., in MPI[5]) and of changes in relevant dynamical features such as vertical wind shear or vorticity. The state of knowledge is summarized in the most recent IPCC report (Solomon *et al.*, 2007). Additionally, a more detailed assessment has been provided by the tropical cyclone research community (WMO, 2006). The following discussion essentially draws from these two documents.

The current generation of coupled climate models greatly under-resolves tropical cyclones. A number of studies exist that analyze tropical cyclone changes in such models. They show largely inconsistent results. For example, Tsutsui (2002) and McDonald *et al.* (2005) reported on an increasing number of intense tropical cyclones under future (warmer) climate conditions. Using a global climate model with about 100 km grid spacing, Sugi *et al.* (2002) found no significant changes in tropical cyclone intensities, but a decrease in global number. Regionally, an increase in frequency was analyzed for the North Atlantic, while a decrease was reported for the North Pacific. Using the same model, but different sea surface temperatures and convection parameterization schemes, Yoshimura *et al.* (2006) also found a decrease in global tropical cyclone frequency and no significant changes for the frequency of intense tropical cyclones. Bengtsson *et al.* (2006) used a coupled atmosphere–ocean model at about 200 km grid spacing and concluded that there was little change in tropical cyclone frequency and intensity in response to global warming, but that regional changes may occur as a result of changes in the location of storm track or its intensity.

Presently, there are two approaches to address the limited capacity of global climate models at simulating tropical cyclone characteristics:

1. *Analysis of large-scale factors relevant for tropical cyclone formation and development.* The climatological conditions under which tropical cyclones develop and occur are well established including warm ocean temperatures, low vertical wind shear, and high values of large-scale relative vorticity in the lower troposphere (Gray, 1968) (see Section 2.3). As models can provide more realistic and physically based scenarios for these large-scale parameters, an assessment of the consequences of such changes on tropical cyclone statistics can be made.

2. *High-resolution simulations.* These comprise either high-resolution simulations with global climate models starting to emerge as a consequence of increased

---

[5] Maximum potential intensity or MPI is a frequently debated measure for tropical cyclone statistics. It basically reflects the thermodynamic aspects relevant for the formation of tropical storms. The key concept is that for any given ocean temperature and atmospheric thermodynamic condition there exists an upper bound for the intensity a tropical cyclone may achieve. This intensity is usually referred to as maximum potential intensity or MPI. Actually, however, only few tropical cyclones reach their MPI, and sensitivity to global warming is only about 3–5% change in MPI per degree warming (WMO, 2006). Knutson and Tuleya (2004) speculated that CO_2-induced tropical cyclone intensity changes are thus unlikely to be detectable in historical records and will probably not be detectable for decades to come.

178 **Past and future changes in wind, wave, and storm surge climates** [Ch. 5

computational capacities (e.g., Oouchi *et al.*, 2006) or high-resolution tropical cyclone models that have been run either for case studies or in idealized experiments under boundary conditions from global experiments forced with increased greenhouse gas concentrations (e.g., Knutson and Tuleya, 2004, 2008).

So far, most existing studies have followed the first approach. Santer *et al.* (2006) showed that over the past several decades sea surface temperatures over most tropical ocean basins have warmed noticeably by several tenths of a degree. As it is most likely that the primary cause of observed global warming is anthropogenic in nature (Solomon *et al.*, 2007), it appears likely that most tropical ocean basins have warmed significantly for the same reason (WMO, 2006). If increasing greenhouse gas concentrations represent the primary cause of these changes, even further increase in the following decades may be expected, improving the conditions for tropical cyclone formation and development (WMO, 2006).

There are, however, a number of other factors that may counteract such a development. Vecchi and Soden (2007) analyzed greenhouse warming–induced changes in large-scale vertical wind shear in an ensemble of 18 state-of-the-art global climate model experiments reflecting the A1B emission scenario.[6] As a robust feature throughout this ensemble, they found substantial increases in vertical wind shear over the tropical North Atlantic and East Pacific as well as over much of the Southern Hemisphere subtropics during the respective local tropical cyclone season (Figure 5.5, see color section). Such an increase may effectively inhibit tropical cyclone development and have deleterious effects on cyclone intensity (Vecchi and Soden, 2007). This was demonstrated by Vecchi and Soden (2007) by analyzing the tropical cyclone genesis potential index—see Equation (2.5)—in more detail. They showed that the various terms contributing to this index will have comparable contributions to overall change, provided their fractional changes are comparable. Figure 5.6 (see color section) shows contributions of the various terms for the Northern Hemisphere tropical cyclone season. It can be inferred that, for the multi-model ensemble, the contribution of the vertical wind-shear term to the overall change in genesis potential is comparable with that of each of the other terms. The net effect of the various contributions on genesis potential is illustrated in Figure 5.7 (see color section). It can be inferred that in the Northern Hemisphere genesis potential increases substantially in the western and central North Pacific, while changes are more modest over the North Atlantic and the eastern North Pacific showing both areas of increases and decreases as well. In the Southern Hemisphere, substantial increases are found over the south Indian Ocean and the western South Pacific. These results indicate that large-scale dynamical and thermodynamic changes are of comparable magnitude. Any assessment of future tropical cyclone activity should thus comprise analyses of both thermodynamic and dynamic effects on tropical cyclone activity.

On an inter-annual time scale ENSO represents the major factor affecting tropical cyclone activity. Over the North Atlantic and the eastern and central North

[6] CO_2 stabilization at 720 ppm by 2100.

Pacific the bulk of tropical cyclones arise from easterly waves. Over the western Pacific tropical cyclone development is associated with a variety of factors comprising easterly waves, monsoon development, and mid-latitude troughs (CCSP, 2008). Any change in the statistics of these phenomena may thus be associated with corresponding changes in tropical cyclone statistics. Uncertainty as to the future development of these phenomena is, however, large (Meehl *et al.*, 2007). Even while there is agreement that projected warming for the 21st century is robust, the magnitude of such warming still has considerable uncertainty. As a consequence, projected changes in tropical cyclone statistics comprise a high degree of uncertainty as they represent the accumulated effect of a number of projected dynamic and thermodynamic changes, the delicate balance of which is eventually responsible for tropical cyclone statistics that will be observed in the future.

Recently, some high-resolution simulations have emerged making use of increased computational power by either running high-resolution global climate models or by using high-resolution tropical cyclone models driven by climate change boundary conditions. Knutson and Tuleya (2004) used mean tropical conditions from an ensemble of nine global models with increased greenhouse gas concentrations to drive a high-resolution tropical cyclone model and concluded that tropical cyclones of greater intensity may emerge in a warmer climate. Oouchi *et al.* (2006) used a high-resolution global climate model driven by the A1B emission scenario to produce two 20-year time slices for present climate conditions and those at the end of the 21st century. They found a decrease in the global number of tropical cyclones and an increase in frequency for those in the North Atlantic. Cyclones of greater intensity were reported only for the North Atlantic and the South Indian Ocean while no significant changes in intensity were reported for other tropical ocean basins. Table 5.1 summarizes the results from medium- to high-resolution global climate models (GCMs). While higher resolution GCMs generally indicate fewer tropical cyclones worldwide, lower resolution models essentially showed no change in global figures. Regional variations among the different studies are substantial, comprising increases and decreases for the same area.

The results discussed may be limited. It is not always clear that simulated changes are larger than the internal variability produced by the model or that natural variability is adequately accounted for. For example, Oouchi *et al.* (2006) analyzed, from a single pair of experiments, two relatively short time slices of 20 years representative of present-day climate and climate at the end of the 21st century. As tropical cyclone activity varies on time scales larger than the simulated period, internal variability could produce changes larger than those associated with radiative forcing.

Summarizing, there is substantial disagreement among the different studies; in particular, between global and regional modeling studies (WMO, 2006). There appears to be some consistency between high-resolution models and theory that some increase in tropical cyclone intensity will occur if the climate continues to warm (WMO, 2006). Increased precipitation rates associated with tropical cyclones appear to be a more robust result (Meehl *et al.*, 2007) simply representing the increased availability of moisture in a warmer atmosphere (WMO, 2006).

180 **Past and future changes in wind, wave, and storm surge climates** [Ch. 5

Table 5.1. Tropical cyclone frequencies in percent of present-day values as simulated by several climate models under enhanced greenhouse gas concentrations. Bold/italic indicates significantly more/fewer cyclones. Table adapted from CCSP (2008).

Reference	Model	Resolution (km)	Experiment	Area			
				Global	N Atlantic	NW Pacific	NE Pacific
Sugi *et al.* (2002)	JMA	120	10 yr time slice, $1 \times CO_2$, $2 \times CO_2$	*66*	**161**	*34*	33
Tsutsui (2002)	NCAR/CCM2	250	10 yr, $1 \times CO_2$, 115 yr, +1% CO_2/yr	102	86	111	91
McDonald *et al.* (2005)	HadAM3	100	15 yr IS95a, 1979–1994, 2082–2097	*94*	*75*	*70*	**180**
Hasegawa and Emori (2005)	CCSR/NIES/ FRCGC	120	5×20 yr, $1 \times CO_2$ 7×20 yr, $2 \times CO_2$			96	
Yoshimura *et al.* (2006)	JMA	120	10 yr, $1 \times CO_2$, $2 \times CO_2$	*85*			
Bengtsson *et al.* (2006)	ECHAM5/OM	200	A1B, 3 runs, 30 yr 20C and 21C	94			
Oouchi *et al.* (2006)	MRI/JMA	20	10 yr A1B, 1982–1993, 2080–2099	*70*	**134**	*62*	*66*
Chauvin *et al.* (2006)	ARPEGE	50	10 yr, SRES-B2		**118**		
Chauvin *et al.* (2006)	ARPEGE	50	10 yr, SRES-A2		*75*		
Bengtsson *et al.* (2007)	ECHAM5	40	20 yr, A1B		*87*	*72*	*107*

5.4 WIND-GENERATED WAVES

5.4.1 Past changes and variability

Long-term changes in large-scale atmospheric circulation coupled with changes in the statistics of tropical and extra-tropical cyclones imply long-term changes in ocean wave statistics. Changes in local wind climatology affect the wind sea, while changes

Sec. 5.4]   **Wind-generated waves**   181

in remote storm statistics affect the swell component of the wave climate. Extra-tropical cyclones dominate the wave climatology at mid-latitudes and higher latitudes, while tropical cyclones dominate at lower latitudes during the warm season.

A number of different data sources exist that have been used to estimate changes in wave climatology. Visual observations and measurements for more than a century are available from voluntary observing ships (VOSs). The availability and quality of these data vary largely over time and space comprising only few observations along the major shipping routes in earlier times. In consequence, data coverage is relatively good, for example, for most parts of the North Atlantic while other areas such as the Southern Ocean are markedly undersampled. Direct or *in situ* measurements can be obtained from buoys or platforms. These are available mostly for a few decades of years and are mostly located relatively close to the coast. Satellite altimetry that can provide a near-global picture has emerged only in recent years. An approach frequently used to assess the wave climate, its variability, and long-term change is to produce hindcasts with numerical wave models driven mostly by reanalyzed atmospheric wind fields (Section 4.4). Such hindcasts are usually limited to typical reanalysis periods (i.e., the last few decades). They may be subject to typical reanalysis problems related to the assessment of long-term changes (Section 4.4.1). In short, the reliability of wave data ranges widely depending on data collection and processing methods.

A reliable global picture of long-term changes in ocean wave climatology is relatively difficult to obtain. Before about 1950, VOS data are available mostly along major shipping routes. After that, near-global coverage may be achieved (Gulev and Grigorieva, 2004). The number of available observations varies between about 15 and 20 per region and month in the late 19th century and several thousands after about 1960, implying time-dependent sampling errors (Gulev and Grigorieva, 2004). In addition, time-dependent fair-weather biases may occur as ships tend to avoid storm conditions, a behavior that may have changed in the course of time in response to changing ship sizes and/or operation practices.

Gulev and Grigorieva (2004) analyzed changes in the wave climate from VOS reports for 1900–2002 along the major shipping routes (i.e., mainly within the North Atlantic and North Pacific). For the North Pacific Gulev and Grigorieva (2004) reported a positive trend in significant wave height (Section 2.4.2) of about 8 cm–10 cm per decade over the last century. For the North Atlantic, they found a negative trend being most pronounced (about 5 cm per decade) within the western part of the North Atlantic storm track. For the period 1950–2002, for which more reliable VOS data with nearly global coverage exist, Gulev and Grigorieva (2004) found positive trends in significant wave height over much of the extra-tropical North Atlantic and North Pacific oceans, the western subtropical South Atlantic, parts of the Indian Ocean, and the east and south China Seas (Figure 5.8, see color section). Large positive trends occurred in the North Atlantic and the North Pacific, negative trends were found for some areas (e.g., around Australia).

Hindcasts represent an alternative approach to assessing long-term changes and variability of the wave climate. One of the first attempts was provided by Kushnir *et al.* (1997) who used a wave model for the North Atlantic driven by operational wind fields from the European Centre for Medium-range Weather Forecast to simulate the 10-year period 1980–1989. They subsequently extended their results statistically by linking hindcast monthly wave height statistics with the observed state of large-scale atmospheric circulation and by using large-scale atmospheric data back to the 1960s to obtain a statistical hindcast of monthly mean wave heights for 1962—1986. They suggested that over that period monthly mean significant wave height had systematically increased in the northeast Atlantic, while it decreased south of about 40°N.

The first multi-decadal wave hindcast for the North Atlantic was provided shortly thereafter by Günther *et al.* (1998) within the European project WASA (WASA-Groupp, 1998). Günther *et al.* (1998) used operational wind fields from the Fleet Numerical Operational Center (FNOC) complemented by winds from the Norwegian Meteorological Office (DNMI) to simulate a period of 40 years from 1955–1994. They concluded that the wave climate in most of the northeast Atlantic and in the North Sea underwent significant variations on time scales of decades. Part of the variability was found to be related to the North Atlantic Oscillation. As a general result, Günther *et al.* (1998) noted an increase in monthly mean significant wave height over the simulated 40 years of about 2 cm–5 cm per decade, and an increase in annual maximum significant wave height of 5 cm–10 cm per year for large parts of the northeast Atlantic.

A *global* wave hindcast covering a period of 15 years (1979–1993) was provided by Sterl *et al.* (1998). They used ERA-15 (Gibson *et al.*, 1996) reanalyzed wind fields (see Section 4.4.3) to force their wave model. In particular, they noted strong upward trends in the North Atlantic and south of Africa but, taking the existing large natural decadal variability into account, the period analyzed appears to be rather short for the assessment of ongoing long-term changes and trends. Sterl *et al.* (1998) concluded that, due to large year-to-year variability, they could not confirm a significant change in significant wave height over the 15-year period.

Wang and Swail (2001) used NCEP/NCAR–reanalyzed wind fields (see Section 4.4.2) to perform a multi-decadal global wave hindcast covering the 40-year period 1958–1997. In their analysis, which focused on the Northern Hemisphere, they noted increasing trends in monthly wave height statistics over this 40-year period. The trends were most pronounced over the northeastern North Atlantic in winter and over the North Pacific in winter and spring. Over the North Atlantic, the increasing trend in the northeastern part was accompanied by a decrease in the subtropical North Atlantic, indicating a corresponding shift in the location of the North Atlantic storm track as a possible explanation. Similar to Kushnir *et al.* (1997), Wang and Swail (2001) used statistical techniques to extend their results backward in time. When this longer period covering almost a century since 1899 was considered, no significant trends in monthly wave height statistics were identified for both oceans. Wang and Swail (2001) noticed, however, strong multi-decadal fluctuations, the time behavior of which for the North Atlantic broadly resembles that of extra-tropical storm activity (Figure 5.3). The existence of such multi-decadal fluctuations may lead to the detection of significant trends when shorter records—in particular, typical reanalysis periods—are considered. The results of Wang and Swail (2001) indicate that care has to be taken in the interpretation of trends obtained from short records (see Section 4.7.3).

The most recent global multi-decadal wave hindcast is the one performed as part of the ERA-40 reanalysis project (see Section 4.4.3). It is described in detail in Sterl and Caires (2005) and comprises 45 years covering the period 1957–2002. Sterl and Caires (2005) noted the importance of the Southern Hemisphere storm track in governing global wave height statistics. In particular, they showed that the dominant mode of global wave height variability (which explains about 15% of the variability) has a pronounced maximum in the Southern Ocean and represents swell propagating into the Pacific and Indian Oceans, while the Atlantic is less affected. As reanalyses are less reliable in the Southern Hemisphere (see Section 4.4.1), this may indicate additional problems in the interpretation of trends obtained from global wave hindcasts driven by reanalyzed wind fields. For the North Atlantic, results were found to be in line with those from earlier studies (e.g., from Günther et al., 1998).

In summary, global and large-scale wave hindcasts to some extent confirm the figures obtained from VOS data—namely, increases in monthly wave height statistics in the North Pacific and the North Atlantic over the last few decades of the 20th century—but hindcasts in combination with statistical reconstructions (e.g., Wang and Swail, 2001) indicate that these increases may be part of existing multi-decadal fluctuations. On a regional and local scale, the figure may or may not deviate from this large-scale picture. Possible reasons for deviations are, for example, local changes in wind direction that are associated with changing fetch or a change in the average duration of high wind speed situations, both with obvious effects on wave statistics. Changing sea ice conditions may also have an impact as less sea ice corresponds to longer fetches and *vice versa*.

There is a large and growing number of studies dealing with long-term changes in wave climate on a local and regional scale, which cannot be treated exhaustively in the context of this book. In the following a few examples are mentioned. Regional multi-decadal wave hindcasts have become a common tool to investigate long-term changes in wave climate on a regional scale. Weisse and Günther (2007) provided a high-resolution (grid spacing of about 5 km) hindcast for the southern North Sea that now covers a period of 60 years (1948–2007). Similar, but somewhat shorter hindcasts exist for the Mediterranean Sea (Musić and Nicković, 2008), the Baltic Sea (Cieślikiewic and Paplińska-Swerpel, 2008), or the waters around New Zealand (Gorman et al., 2003), to name a few. Other studies are related to the analysis of buoy data. Bromirski et al. (2005) analyzed wave spectral energy variability in three different frequency bins representing wind sea, swell, and mixed conditions using data from a number of buoys in the northeast Pacific. They found that the spatial pattern of variability is similar for all three frequency bins, and that an increasing trend over the period 1981–2003 could be identified mainly in the swell component, an indication that storm activity in the northeast Pacific had increased over the same period. Komar and Allan (2007, 2008) used data from deep-water buoys off the U.S. East Coast over a range of latitudes, where wave climate is affected by both tropical and extra-tropical cyclones. Providing a separate analysis for winter, where wave climate is dominated by extra-tropical cyclones, and summer, where wave climate is

184 **Past and future changes in wind, wave, and storm surge climates** [Ch. 5

dominated by tropical cyclones, they found indications that only the latter has roughened when two periods with relatively low (1977–1990) hurricane activity and relatively high (1996–2005) hurricane activity were compared.

Summarizing, long-term changes and variability in wave climate are strongly linked to changes in atmospheric circulation and to variations in the statistics of tropical and extra-tropical cyclones. In accordance with fluctuations in these parameters, there appears to be considerable multi-decadal variability in wave climate with indications that large-scale wave climate became rougher for about four decades at the end of the last century for most Northern Hemisphere oceans. There are also indications that this intensification may reflect large-scale multi-decadal natural variability as analysis of century-long records so far does not reveal any significant changes. Regionally, changes may be different and modulated by changing fetch, duration, etc., which in turn may reflect changes in the prevailing direction of strong winds, sea ice conditions, or changes in local bathymetry, etc.

5.4.2 Future changes

Future changes in ocean wave climate may occur from changes in large-scale atmospheric circulation, changes in the statistics of tropical and extra-tropical cyclones, or changes in mean sea ice conditions that may result in modified fetch conditions. Locally and in particular near the coasts, changes in water depth resulting from erosion, and water work such as dredging, etc. may play a role.

While there is presently a large number of studies on possible future changes in large-scale atmospheric circulation or storm statistics, the number of studies explicitly engaged in possible future changes in ocean waves appears rather limited. The majority of studies are based on statistical downscaling approaches (Section 4.5) that link changes in large-scale atmospheric circulation to changes in the statistics of significant wave heights (e.g., Wang *et al.*, 2004; Caires *et al.*, 2006). Such studies usually use atmospheric reanalysis data and related wave hindcasts to train the statistical model. Subsequently, these statistical models are applied to the output of climate models simulating future climate conditions. In this way, inferences about large-scale changes in ocean wave climate are obtained.

Wang *et al.* (2004) used a statistical approach to provide projections for future significant wave height conditions in the North Atlantic. They trained their model using NCEP/NCAR–reanalyzed sea level pressures and a wave hindcast driven by NCEP/NCAR–reanalyzed wind fields (Section 4.4.2). Subsequently, the statistical model was applied to output from the CGCM2 (Canadian Centre for Climate Modelling and Analysis) coupled atmosphere–ocean model for three realizations of three different greenhouse gas emission scenarios (i.e., nine simulations in total). Towards the end of the 21st century, Wang *et al.* (2004) found noticeable increases in monthly wave height statistics in the northeast Atlantic under all three emission scenarios, with stronger responses associated with stronger emission scenarios. Increases were also found for the southwest North Atlantic, while decreases occurred in mid-latitudes. All changes were in the order of about 5% of the baseline climate. For the North Pacific, a similar analysis was presented in Wang and Swail (2006b).

A multi-model and global, but otherwise similar approach was presented by Wang and Swail (2006a). In particular, they used output from three *different* coupled climate models, one of which was again the CGCM2. In total, the ensemble of Wang and Swail (2006a) thus comprised15 members.[7] The results of analyses for the A2 and B2 emission scenarios are shown in Figures 5.9 and 5.10 (see color section for both). For the A2 scenario, strongest changes generally occur in the Southern Ocean. They would be consistent with the poleward movement of the mid-latitude storm track, with corresponding wave height increases at high latitudes, and decreases at mid-latitudes. For the B2 scenario, a similar but somewhat weaker response was found. The Northern Hemisphere picture is broadly consistent with that described in Wang *et al.* (2004) and Wang and Swail (2006b). Note, however, that the increase in winter (January–March) mean significant wave height in the northeast Atlantic in the A2 scenario is replaced by a decrease in the B2 scenario, indicating considerable uncertainty—in particular, on smaller scales.

An important point is to quantify how much of the uncertainty can be attributed to variations among the different scenarios—that is, to our uncertainty about future socio-economic and social developments—and how much may be accounted for by the use of different climate models. The latter reflects our uncertainty in knowledge and understanding of the climate system (Section 5.1). For their experiments, Wang and Swail (2006a) found that the differences in wave height response among the different *forcing* scenarios were much smaller than those among the different *models* used to simulate the changes. This indicates that either the different emission scenarios reflect a relatively narrow band of possible future conditions, or that model uncertainties are indeed overwhelming. In any case, model-inherent uncertainties are presently the dominant source of uncertainty.

Similar results were found by other authors for other model/scenario combinations. For example, Grabemann and Weisse (2008) used near-surface marine wind fields from two different coupled climate models[8] for two different (A2, B2) emission scenarios to run a high-resolution wave model for the North Sea for two time slices, 1960–1990 and 2071–2100. Figure 5.11 shows the climate change signal in terms of 99-percentiles of significant wave height. Changes were generally smaller/larger when downscaled wind fields from the HadAM3/ECHAM5 model were used. The largest change in 99-percentile significant wave height obtained by using HadAM3 wind fields was about 20 cm in the southern North Sea, while it was about 80 cm and located in the Skagerrak when ECHAM5 wind fields were used. For ECHAM5 wind fields, the response is generally stronger for the A2 scenario. For HadAM3 wind fields, largest wave height changes were obtained for the B2 scenario. ECHAM5-driven wave height changes are generally larger than those obtained for HadAM3, even for the southern North Sea and the B2 scenario where HadAM3 showed strongest changes. This reveals that uncertainties—in particular, on smaller scales—may be considerable. Grabemann and Weisse (2008) showed that the primary source of uncertainty are differences in regional wind fields that may be a result

---

[7] 2 models × 3 scenarios × 1 realization + 1 model × 3 scenarios × 3 realizations.
[8] Note that winds were additionally downscaled by a regional climate model.

186 Past and future changes in wind, wave, and storm surge climates   [Ch. 5

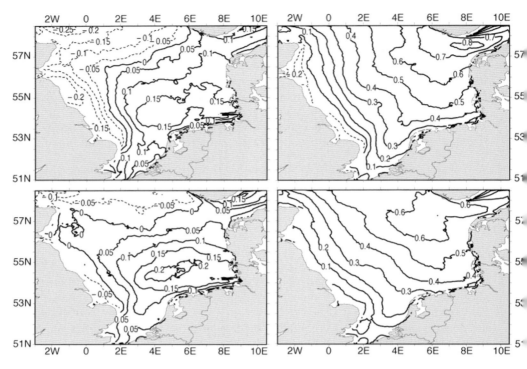

Figure 5.11. Climate change signals in meters for long-term 99-percentile significant wave height from wave model simulations driven with downscaled HadAM3 (left) and ECHAM5 (right) wind fields for the A2 (top) and the B2 (bottom) emission scenario. Redrawn from Grabemann and Weisse (2008).

of regional differences in the climate change response and/or of different model sensitivities.

The results described so far indicate that larger multi-model ensembles covering a broad range of emission scenarios are required. Additionally, dynamical approaches would be an advantage to better assess the expected range of uncertainties. While global studies may provide a broad-scale picture, high-resolution regional studies are an advantage to reach conclusions relevant for local planning and response. Such studies are beginning to emerge. Early attempts were made by the WASA-Group (1998) and by Kaas *et al.* (2001) who simulated relatively short periods of 5-year and 30-year changing ocean wave conditions for the North Atlantic, using the time slice technique. Later a more comprehensive approach was provided by Debernhard *et al.* (2002) for northern seas. More recently, multi-model ensembles begin to emerge, although the size of the ensembles so far is still rather limited (e.g., Grabemann and Weisse, 2008; Debernhard *et al.*, 2008). Similar approaches on a global scale are, to our knowledge, not yet available.

5.5 TIDES, STORM SURGES, AND MEAN SEA LEVEL

Water levels, or the height of the sea surface, have changed in a number of different ways during the recent past and will continue to do so in the future. Changes comprise variations in the tidal, mean sea level, or storm surge components (Section 2.5). Existing publications so far mostly deal with changing mean sea level. Changing tides and storm surges are less often considered. In the following, we will review past and future changes in the three components individually. We concentrate on time scales of decades up to a century. Longer time scales are not considered.

5.5.1 Mean sea level

5.5.1.1 *Past changes and variability*

There are a number of different processes that contribute to changes and variations in mean sea level (see Section 2.5.3). On decadal to century time scales, there are basically two processes that contribute significantly to observed changes: thermal expansion/contraction of ocean water in response to ocean warming/cooling and exchange of water with land-based reservoirs such as glaciers, ice caps, ice sheets, etc. (Bindoff *et al.*, 2007). Both processes alter the *volume* of the oceans and thus *global* mean sea level. On a regional scale, processes that *redistribute* mean sea level with nearly zero global change may significantly modify the global mean picture. For example, long-term changes in ocean circulation will lead to a redistribution of mass and thus of regional mean sea level. Caused by the inverse barometric effect (Section 2.5.1), long-term changes in the spatial distribution of atmospheric surface pressure would have similar consequences. In addition, vertical land movements caused by glacial isostatic adjustment (GIA), plate tectonics, etc. may be relevant on a regional scale, but may, in contrast, also have an impact on global mean sea level as they change the volume of ocean basins.

There are in principle two different data sources from which long-term changes in mean sea level can be determined: tide gauges and satellite altimetry. Tide gauges are available for longer periods, but their spatial coverage is not optimal. Satellite altimetry provides near-global (between 66°S and 66°N) coverage, but became available only from 1993 onwards. Both data sources have their limitations and sources of uncertainty. A more detailed discussion can be found in Section 2.5.3 or in Bindoff *et al.* (2007).

Over the 20th century and estimated from tide gauge data, global mean sea level has increased at a rate of about $1.7 \, \text{mm} \, \text{yr}^{-1}$ (Bindoff *et al.*, 2007). For the period 1993–2006, satellite altimetry provided a rate of about $3.1 \, \text{mm} \, \text{yr}^{-1}$ (Church *et al.*, 2008), which was confirmed by tide gauge data until 1999. After 1999, a noticeable disagreement between estimates based on tide gauge and satellite data occurred, careful analysis of which is urgently needed (Domingues *et al.*, 2008). Figure 5.12 (see color section) shows the development of global mean sea level over the past century as obtained from tide gauges and from satellites over the past decade (Church *et al.*, 2008). The authors additionally provided estimates of sea level trends over

20-year periods, with the start date of each period incremented by one year. Analysis indicates significant variability in the rate of sea level rise over the last century with rates being smaller than about $1\,\text{mm}\,\text{yr}^{-1}$ before about 1930. Between 1930 and 1950 rates increased to about $2\,\text{mm}\,\text{yr}^{-1}$ to $2.5\,\text{mm}\,\text{yr}^{-1}$. Between 1963 and 1991 some reduction occurred which was attributed by some authors to a series of volcanic eruptions that caused ocean cooling and thus contraction (Church *et al.*, 2008). After that period, rates increased again with the latest values being the highest on record so far and close to the estimates derived from satellite altimetry. Regionally, the spatial pattern of the global mean increase is not uniform. As an example, Figure 5.13 (see color section) shows the spatial distribution of the rates of sea level change 1993–2003 derived from satellite altimetry. It can be inferred that over that period largest increases in mean sea level were observed in the western Pacific, while changes in other areas are smaller. Because of the short period considered, the pattern basically reflects climate variability associated with the El Niño–Southern Oscillation phenomenon (Church *et al.*, 2008). On longer time scales, the pattern is still not uniform, but short-term climate variability has a weaker impact (Church *et al.*, 2004).

There is considerable uncertainty in our understanding of how sea level has changed on decadal and longer time scales and of the contributions made by various processes (Church *et al.*, 2008). Ocean thermal expansion is thought be the largest contributor during the 20th century. Between 1993 and 2003, ocean thermal expansion is estimated to have contributed about $1.2\,\text{mm}\,\text{yr}^{-1}$ to $1.6\,\text{mm}\,\text{yr}^{-1}$ to observed global mean sea level rise (Antonov *et al.*, 2005; Ishii *et al.*, 2006; Willis *et al.*, 2004). These estimates are based to some extent on data from expendable bathy-thermographs (XBTs), devices that measure ocean temperature as a function of depth where depth is determined by *a priori* knowledge on the rate at which the sensors are sinking. Recently, Domingues *et al.* (2008) showed that an apparent time-dependent bias in XBT sinking rates[9] leads to a time-dependent warm bias in XBT data. Correcting for this bias, Domingues *et al.* (2008) provided an updated estimate of $0.8\,\text{mm}\,\text{yr}^{-1}$ for ocean thermal expansion for the period 1993–2003, considerably smaller than earlier numbers. When somewhat longer periods are considered, estimates of about $0.3\,\text{mm}\,\text{yr}^{-1}$ for 1955–2003 (Antonov *et al.*, 2005), $0.4\,\text{mm}\,\text{yr}^{-1}$ for 1961–2003 (Bindoff *et al.*, 2007), and $0.5\,\text{mm}\,\text{yr}^{-1}$ for 1960–2003 (Domingues *et al.*, 2008) are obtained for the contribution of ocean thermal expansion to observed mean sea level rise.

The second largest contribution to 20th-century sea level rise is thought to arise from the melting of glaciers and ice caps, but excluding the Greenland and Antarctic Ice Sheets (Church *et al.*, 2008). Kaser *et al.* (2006) estimated the contribution to be about $0.4\,\text{mm}\,\text{yr}^{-1}$ for 1961–1990. Bindoff *et al.* (2007) provided estimates of about $0.5\,\text{mm}\,\text{yr}^{-1}$ and $0.77\,\text{mm}\,\text{yr}^{-1}$ for 1961–2003 and 1993–2003, respectively. The large ice sheets of Greenland and Antarctica are thought to have provided only minor contributions during the 20th century, but are potentially the largest contributer in the future (Church *et al.*, 2008) comprising the largest uncertainty. Changes in terrestrial storage—comprising changes in groundwater, soil moisture, or snow

---

[9] Probably caused by changes in manufacture (Church *et al.*, 2008).

Table 5.2. Estimates of contributions by various processes to global mean sea level budget. Adapted from Bindoff *et al.* (2007).

Source	Global mean sea level rise (mm yr^{-1})	
	1961–2003	1993–2003
Thermal expansion	0.42 ± 0.12	1.60 ± 0.50^a
Glaciers and ice caps	0.50 ± 0.18	0.77 ± 0.22
Greenland Ice Sheet	0.05 ± 0.12	0.21 ± 0.07
Antarctic Ice Sheet	0.14 ± 0.41	0.21 ± 0.35
Total	$\mathbf{1.10 \pm 0.50}$	$\mathbf{2.80 \pm 0.70}$
Observed	$\mathbf{1.80 \pm 0.50}$	$\mathbf{3.10 \pm 0.70}$}
Difference	*0.70 ± 0.70*	*0.30 ± 1.00*

[a] Note that this estimate was substantially challenged by Domingues *et al.* (2008) who estimated a thermal expansion contribution of 0.8 mm yr^{-1} after removal of a time-dependent bias in XBT data.

cover—may also contribute to sea level changes, but the terms are poorly known (Church *et al.*, 2008).

Table 5.2 adapted from Bindoff *et al.* (2007) provides estimates of the sea level budget for two periods 1961–2003 and 1993–2003. For the 1961–2003 period it is obvious that the budget is not closed; that is, the estimate obtained from estimates of individual contributions deviates noticeably from the observed trend. For 1993–2003 the budget appeared to be closed within error bounds; however, it is now known that thermal expansion was overestimated for that period (Domingues *et al.*, 2008). When the error is corrected for, a difference comparable with that for 1963–2003 is obtained. This indicates that there is still insufficient understanding as to how sea level has changed and what the individual contributions from various processes had been. Closing the sea level budget remains an area of active research.

5.5.1.2 *Future changes*

Presently the most comprehensive assessment of possible future changes in mean sea level is provided by the Fourth IPCC Assessment Report (Meehl *et al.*, 2007).[10] It is

---

[10] Remarkably, the assessment on sea level rise by the Fourth Assessment Report of the IPCC meets considerable skepticism. According to a survey by Bray in 2008, about 40% of respondents agreed to the statement "The IPCC tends to underestimate the magnitude of future changes of sea level" (Bray, pers. commun.). 13% considered the assessment to be an overestimate, while only 47% considered the IPCC representation to be accurate.

190 **Past and future changes in wind, wave, and storm surge climates** [Ch. 5

based on detailed assessment of ocean thermal expansion from climate models, melting mountain glaciers by scaling observations to projected atmospheric temperature rise, ice sheet mass balance changes and dynamic response from ice sheet models, and extrapolation of recent observations. In this way, an increase in global mean sea level from 1980–1999 to 2090–2099 between 18 cm and 59 cm is projected (Figure 5.14, see color section). In all scenarios, the largest contribution is obtained from thermal expansion (10–41 cm) while mountain glaciers and ice caps still provide the second largest contribution (7–17 cm) to projected global mean sea level rise. In total, the contributions from the Greenland and Antarctic Ice Sheets are small, partly because increased accumulation over Antarctica to some extent offsets increased ablation from elsewhere.

There is, however, increasing concern regarding the stability of the ice sheets in the near future. In the Fourth IPCC Assessment, this is accounted for by an additional temperature-dependent contribution of 10 cm to 20 cm, referred to as the scaled-up ice sheet dynamical imbalance (Figure 5.14). Note that this contribution is not accounted for in the total estimate in Figure 5.14. When accounted for, the total range for projected global mean sea level change towards the end of the 21st century becomes 18 cm to 79 cm. Meehl *et al.* (2007) noted that the uncertainty in these estimates is considerable and that even larger values cannot be excluded. According to the IPCC, the present understanding of the processes involved is, however, too limited to provide best estimates or an upper bound.

The IPCC has made an attempt to account for the uncertainties related to ice sheet response. However, there remains concern in the scientific literature that ice sheet contributions to global mean sea level rise may have been underestimated. This is partly caused by the finding that observed sea level changes from 1990 to the present have been larger than that projected by the IPCC's central value for the same period (Church *et al.*, 2008). Rahmstorf (2007) developed a simple statistical model relating 20th-century temperature change with global mean sea level and applied this relation to 21st-century temperature changes. He provided estimates of global mean sea level change that were considerably larger than those provided by the IPCC's Fourth Assessment Report (Meehl *et al.*, 2007). The approach was, however, contested on both physical and statistical grounds (Schmith *et al.*, 2007; Holgate *et al.*, 2007; von Storch *et al.*, 2008b).

On the regional scale, mean sea level changes may deviate significantly from the global mean picture. The processes responsible for such deviations comprise

1. *Self-gravitational effects of shrinking ice masses.*[11] Large land-based ice masses exert a gravitational pull on the surrounding ocean. As a consequence, mean sea level in the vicinity of the ice mass is increased. When the ice mass shrinks, the gravitational pull is reduced and the freshwater released is not distributed evenly over global oceans. Actually, mean sea level will drop in the immediate vicinity of

---

[11] Although important, the effect was overlooked by the oceanographic community for a long time, even though it had been detected and analyzed in the late 19th century. For a historical account see Vermeersen and Schotman (2009).

a shrinking ice mass (the so-called near-field response), and in the far field will rise much more than the global mean sea level equivalent corresponding to released freshwater. In an intermediate zone, mean sea level actually rises, but less than the global mean sea level equivalent of the freshwater released. Moreover, the solid Earth will deform under shifting loads resulting in changes in the gravity field with a corresponding redistribution of seawater and the vertical position of the land. For mass changes in the Greenland and Antarctic Ice Sheets as well as for shrinking mountain glaciers the effects have been estimated by Mitrovica *et al.* (2001) and are illustrated in Figure 5.15 (see color section). Shrinking ice masses exhibit characteristic spatial patterns in regional mean sea level change usually referred to as fingerprints. Values smaller than zero indicate regions where sea level actually drops when the ice masses considered shrink. Values between 0 and 1 indicate that the increase in regional mean sea level will be smaller, values larger than 1 that the increase will be larger than the global mean sea level equivalent corresponding to the amount of freshwater released by a shrinking ice mass.

2. *Local expansion and redistribution of mass.* There may be regional differences in ocean temperature changes caused by a combination of changes in differential heating and temperature changes due to variations in ocean circulation. Moreover, ocean circulation changes will lead to barotropic adjustments and consequently to a redistribution of mass resulting in regionally different sea level changes. Estimates of these changes are usually obtained from climate models (Meehl *et al.*, 2007). However, the regional patterns of sea level change derived from the different models participating in the IPCC's Fourth Assessment Ensemble (Meehl *et al.*, 2007) are not similar. The largest pattern correlation found between any pair of models is about 0.75, but only 25% of the correlations were larger than 0.5 (Meehl *et al.*, 2007). There are just a few areas where the model ensemble mean change exceeds inter-model variability (standard deviation). These are primarily the Southern Ocean where smaller-than-global-average sea level rises are projected, and the South Atlantic and the south Indian Ocean where larger changes occurred.

3. *Redistribution of atmospheric pressure loading.* Caused by the inverse barometric effect (Section 2.5.1), low-frequency changes in atmospheric pressure loading may lead to regional difference in mean sea level. Using a coupled climate model under a climate change scenario, Stammer and Hüttemann (2008) estimated that the changes caused by these effects could be as large as 10%–20% of anticipated global mean sea level rise over the next 100 years. Note that this effect is usually not taken into account in model-based studies on climate change–related sea level changes (Stammer and Hüttemann, 2008).

4. *Local land movements.* These may result from glacial isostatic adjustment (GIA), plate tectonics, or other local processes. The latter may comprise, for example, subsidence due to drainage and/or ground water, oil or gas extraction, etc. Tide gauges can, sometimes significantly, be affected by local land movements. Additional evidence is then needed to propely intepret their data.

192 **Past and future changes in wind, wave, and storm surge climates** [Ch. 5

5.5.2 Storm surges

5.5.2.1 Past changes and variability

The probability of flood risk in coastal areas is related to the distribution of extreme sea levels. Locally, changes in highest water (sea) levels occur mainly as a result of two effects: local changes in mean sea level—that is, the present extreme levels or thresholds will be exceeded more/less frequently—and changes in the statistics of the storm surge component of the water level—that is, the frequency, height, or duration of surge events.

An approach for separating both effects was suggested by de Ronde (WASA-Group, 1998). The approach is based on the assumption that mean sea level changes will shift the entire frequency distribution of local water level heights towards higher or lower values; that is, the signal will be coherently visible in both mean and extreme sea levels. On the other hand, any change in the statistics of storm surges will have only minor effects on the mean, but will be clearly visible in the extreme sea level time series. De Ronde therefore suggested removing the coherent part of the signal from time series of extreme water levels. This can be done by subtracting annual means from annual upper percentiles (such as 99-percentiles). The difference then represents a proxy for storm-related water level variations. Mean sea level variations, on the other hand, can be considered by analyzing the coherent part from both time series.

The approach was applied by von Storch and Reichardt (1997) to water levels in Cuxhaven on the North Sea coast of Germany. Figure 4.12 (see color section) shows an update of this analysis for 1843–2006. The annual mean time series clearly shows an increasing trend over the more-than-150-year-long record. A similar increase is obtained when annual 99-percentiles are considered (not shown). When the difference between both time series is analyzed, no long-term change is obtained but pronounced inter-annual and decadal fluctuations that broadly resemble the known variations of storm activity in the area.

A similar analysis was provided by Woodworth and Blackman (2004) for a set of 141 tide gauges starting mostly around 1975 and distributed unevenly over the world oceans. The results of this analysis are shown in Figure 5.16 (see color section). When raw 99-percentiles were considered, a number of gauges showed significant increases in extreme sea level since 1975 mostly along the Atlantic coast of the United States, within the central and South Pacific, and along parts of the coasts of Japan and China. When the mean sea level contribution to this change is removed, the signals along the U.S. East Coast and in Asia mostly dissappear, with some significant trends remaining mostly in the central and South Pacific. Woodworth and Blackman (2004) note that the time series considered is relatively short and that long-term tidal components (Section 2.5.2) may have an influence on estimated trends. When the latter is removed, most of the significant changes in the Pacific go as well, indicating that, within the data set and the period considered, trends in extreme sea levels are primarily a result of coherent changes in mean sea level and of artifacts related to short-term trends in long-term astronomical tides (Woodworth and Blackman, 2004). In contrast, the storm surge component does not seem to show any long-term trend. Woodworth and Blackman (2004) concluded that variations in mean and extreme sea

Sec. 5.5]   **Tides, storm surges, and mean sea level**   193

level have been largely coherent throughout their analysis period, indicating that they have been produced by the same type of forcing.

Worldwide, there is only a small number of long tide gauge records that allow analysis of decadal and longer changes in the storm surge component. Zhang *et al.* (2000) analyzed changes in extreme sea levels since 1900 at 10 tide gauges along the U.S. East Coast. In agreement with Woodworth and Blackman (2004) they conclude that the observed rise in extreme sea levels since about 1900 closely follows mean sea level trends and that there are no long-term changes in storm-related components. Bromirski *et al.* (2003) analyzed hourly tide gauge data from San Francisco for the period 1858–2000. They concluded that the storm surge component did not change substantially over the analysis period. For the Liverpool tide gauge, Woodworth and Blackman (2002) provided an analysis of several storm surge measures back to 1768. They noted that surges showed considerable fluctuations on inter-annual and decadal time scales consistent with observed variations in storm activity, but revealed no long-term trend. A similar conclusion was obtained by von Storch and Reichardt (1997) for Cuxhaven and from an update of their analysis for the period 1843–2006 (Figure 4.12, see color section).

5.5.2.2   *Future changes*

Apart from mean sea level changes, extreme sea level changes may arise from changes in storm surge statistics. These, in turn, will change, when the statistics of storms (path, wind direction, duration, frequency, intensity, etc.) in an area change. Extreme sea levels are thus affected by local meteorology and, moreover, by small-scale geographical features such as the shape of the coastline or bathymetry. Studies on changing storm surge statistics therefore generally have a regional focus and results may not be generalized for other regions or the globe.

As for ocean waves, statistical (e.g., von Storch and Reichardt, 1997) and dynamical (e.g., Kauker and Langenberg, 2000; Langenberg *et al.*, 1999; Lowe and Gregory, 2005; Woth, 2005) downscaling approaches are used. The latter usually apply more or less sophisticated regional tide–surge models driven by near-surface wind and pressure fields obtained from global or regional climate change simulations.

A number of studies exist for the northwest European shelf with mostly consistent results, pointing towards an increase of wind-related effects mainly in the southeastern part of the North Sea on the order of a few decimeters by the end of the 21st century, and towards little changes elsewhere. Some of these studies used statistical (e.g., von Storch and Reichardt, 1997) and others dynamical downscaling approaches (e.g., Lowe and Gregory, 2005; Woth *et al.*, 2006; Sterl *et al.*, 2008). The results of different studies are not easily comparable as different time horizons (e.g., time of doubling of CO_2 concentrations, or the end of the 21st century), different scenarios, and different parameters characterizing the storm surge climate (e.g., high percentiles or extreme values) were used. In general, the patterns and the magnitude of changes were, however, mostly consistent. Not in all studies were all changes found to be detectable; that is, some were within the range of natural variability.

194 Past and future changes in wind, wave, and storm surge climates [Ch. 5

Another area where climate change–related effects on storm surge statistics have been studied extensively is the coastline of Australia. For parts of the northern coastline, which is affected by tropical storm surges, McInnes *et al.* (2003) reported increases in the area inundated by storms when estimates of climate change to 2050 comprising both changes in storm statistics and mean sea level rise are considered. For coasts along the south of Australia, where surges are driven by extra-tropical storms, McInnes and Hubbert (2003) and McInnes *et al.* (2005) reported a mixed picture from analysis of an ensemble of climate change simulations some of which projected increase while others showed decrease in extreme sea level statistics. The largest contribution was generally obtained from an increase in mean sea level, rather than from the wind-driven component.

Analysis of changing storm surge statistics for other areas is less comprehensive. Some studies exist, for example, for the Bay of Bengal, an area that is heavily threatened by the impact of storm surges from tropical cyclones. Earlier studies (Flather and Khandker, 1993) analyzed the impact of changing mean sea level on the height of the most severe storm surges and reported a regionally mixed picture with some areas in the Bay showing increases and other decreases. Unnikrishnan *et al.* (2006) used a tide–surge model for the Bay driven by the output of one climate change scenario. As this climate model simulation showed increased tropical cyclone frequency in the area, the authors noticed a corresponding increase in the frequency of high surges.

All projections of possible future storm surge heights depend significantly on expectations about the rise of regional mean sea level. As outlined above, this is still under debate.

5.5.3 Tides

Tides are driven by the gravitational forces of the Moon and Sun, the movements of which are extremely regular and stable on the time scales considered in this book. The open ocean response to these forces—that is, open ocean tides—may change additionally when depth and the shape of ocean basins are modified. Such changes occur mostly slowly over geological time scales and, as a consequence, open ocean tides are probably relatively stable over time scales of decades up to a few centuries. Pugh (2004) listed some evidence from analyses of long tide gauge records that are well connected to the open ocean. The results indicate that open ocean tidal ranges have probably changed less than about 2% over the last 100–200 years.

Coastal tides are the result of tidal energy propagating into the area from the deep oceans. In coastal areas and on a regional and local scale, natural processes and coastal engineering may significantly modify the shape of the coastline, water depth, etc. As a consequence, significant changes in tidal range may occur. In London, about 80 km upstream in the River Thames, tidal ranges have increased by about 70 cm over the last century (Pugh, 2004). A similar situation is found in Hamburg, about 100 km upstream of the River Elbe, where tidal high waters have increased by about 50 cm to 60 cm and tidal low waters decreased by about 100 cm since 1962 mainly as a result of major engineering water works (von Storch *et al.*, 2008a). As a result, tidal ranges in Hamburg significantly increased over the last few decades.

When global mean sea level is rising, tidal wavelengths will increase and tidal patterns will be stretched (Pugh, 2004). However, except for very shallow water where the change in water depth caused by changing mean sea level may be significant, tidal changes will probably be small. Figure 5.17 (see color section) shows an example of a sensitivity study for the southern North Sea and the Elbe estuary. It can be inferred that, for the open North Sea, changes are generally small, but may be more substantial near the coast and within the Elbe estuary. Generally, changes in tidal patterns and amplitudes are stronger for larger mean sea level changes.

5.6 SUMMARY

Extra-tropical cyclone statistics—that is, their average frequency, intensity, path, etc.—have varied substantially on time scales of decades and longer. Over the last few decades some shift in extra-tropical storm activity towards higher latitudes seems to have occurred in both hemispheres. However, when longer periods up to a century and more are considered, changes appear to fall mostly within the range of observed natural variability. From the few available long-term proxy records, there is no indication that extra-tropical storm climate has worsened substantially over the last century. Ocean wave statistics are closely related to storm statistics. Their analysis basically confirms the picture derived from atmospheric data.

Records for tropical cyclone statistics are extremely short for most basins. Moreover, inhomogeneities caused by major changes in the observing system may hamper the analysis of long-term changes. Over the last few decades, tropical cyclone activity appeared to be relatively constant over the Northern Hemisphere, while regionally increases and decreases may have been observed. The longest record exists for North Atlantic hurricanes. Hurricane activity correlates well with sea surface temperature in the area with some authors arguing the behavior to be mostly oscillatory, while others attribute the changes to changes in radiative forcing caused by natural variations and anthropogenic climate warming. This issue and interpretation of the hurricane record are still heavily debated.

Water level changes may be caused by a number of different factors. Global mean sea level was estimated to increase by about $1.8 \, \text{mm} \, \text{yr}^{-1}$ over the last century. Regionally, this increase has not been uniform with larger values being observed for the western Pacific between 1993 and 2003. Most of these variations have been associated with large-scale climate variability such as El Niño. There remain considerable uncertainties regarding our understanding of mean sea level changes. In particular, the sea level budget for the last few decades is far from being closed. When contributions from individual source terms in the sea level equation are estimated, the sum of these terms is considerably smaller than the observed increase. Extreme water levels have changed mostly in coherence with mean sea level changes. There is scant evidence for a systematic change in the storm surge component, other than multi-decadal variability associated with varying storm activity.

Estimates for future changes in wind, wave, and sea level statistics are described for scenarios of anthropogenic climate change. For extra-tropical cyclones, the Fourth IPCC Assessment Report notes the poleward shift in storm activity by several degrees latitude in both hemispheres being a result emerging more consistently across different climate models and studies. A tendency towards fewer, but more intense mid-latitude cyclones, which was noted by earlier IPCC reports, is less emphasized.

When potential future changes in tropical cyclone statistics are discussed, there remains substantial disagreement among different studies. A statement provided by the tropical cyclone research community (WMO, 2006) notes that there is some agreement among models and the theory that tropical cyclones may intensify when the climate continues to warm. As a more robust result, however, a possible increase in tropical cyclone precipitation is emphasized.

Expected changes in ocean wave heights broadly correspond to expected changes in storm statistics. On a regional scale, changes in prevailing wind direction, fetches, or the duration of storms may play a role. Changing sea ice conditions may also modify fetch and may thus have an impact on wave statistics. As a consequence, uncertainty in regional projections is considerable and larger ensembles are required to provide a more consistent picture.

Mean sea level is expected to rise between about 18 cm and 59 cm towards the end of the 21st century. When the additional contribution from potentially reduced ice sheet stability is accounted for, a range between 18 cm and 79 cm is provided by the Fourth IPCC Assessment Report. These numbers are still debated in the scientific literature. At the regional scale, a number of processes may lead to deviations from the global figure. Examples are self-gravitational effects from shrinking ice sheets, vertical land motion, or redistribution of mass by ocean circulation changes. There is no general statement regarding global changes in storm surge statistics. At the regional scale, these may change in correspondence to changes in local storm statistics. Tidal changes may, on the time scales considered, arise from changes in the shape of the coastline and/or water depth. Such changes may be caused naturally or by water works and will have an impact only locally in very shallow regions. Mean sea level rise may also change tidal dynamics. Again, the effect will be mostly visible in shallow coastal areas and in estuaries.

5.7 REFERENCES

Alexandersson, H.; T. Schmith; K. Iden; and H. Tuomenvirta (1998). Long-term variations of the storm climate over NW Europe. *Global Atmos. Oc. System*, **6**, 97–120.

Alexandersson, H.; H. Tuomenvirta; T. Schmith; and K. Iden (2000). Trends of storms in NW Europe derived from an updated pressure data set. *Climate Res.*, **14**, 71–73.

Angel, J.; and S. Isard (1998). The frequency and intensity of Great Lake cyclones. *J. Climate*, **11**, 61–71.

Antonov, J.; S. Levitus; and T. Boyer (2005). Thermosteric sea level rise, 1955–2003. *Geophys. Res. Lett.*, **32**, L12602, doi: 10.1029/2005GL023112.

Bärring, L.; and H. von Storch (2004). Scandinavian storminess since about 1800. *Geophys. Res. Lett.*, **31**, L20202, doi: 10.1029/2004GL020441.

Bell, G.; M. Halpert; R. Schnell; R. Higgins; J. Lawrimore; V. Kousky; R. Tinker; W. Thiaw; M. Chelliah; and A. Artusa (2000). Climate assessment for 1999. *Bull. Amer. Meteorol. Soc.*, **81**, 1328.

Bengtsson, L.; S. Hagemann; and K. Hodges (2004). Can climate trends be calculated from reanalysis data? *J. Geophys. Res.*, **109**, doi: 10.1029/2004JD004536.

Bengtsson, L.; K. Hodges; and E. Roeckner (2006). Storm tracks and climate change. *J. Climate*, **19**, 3518–3543.

Bengtsson, L.; K. Hodges; M. Esch; N. Keenlyside; L. Kornblueh; J. Luo; and T. Yamagata (2007). How may tropical cyclones change in a warmer climate? *Tellus*, **59**, 539–561.

Bengtsson, L.; K. Hodges; and N. Keenlyside (2009). Will extra-tropical storms intensify in a warmer climate? *J. Climate*, **22**, 2276–2301.

Bindoff, N.; J. Willebrand; V. Artale; A. Cazenave; J. Gregory; S. Gulev; K. Hanawa; C. Le Quéré; S. Levitus; Y. Nojiri *et al.* (2007). Observations: Oceanic climate change and sea level. *Climate Change 2007: The Physical Science Basis. Contribution of Working Group I to the Fourth Assessment Report of the Intergovernmental Panel on Climate Change.* Cambridge University Press, Cambridge, U.K. ISBN 978-0-521-88009-1.

Bromirski, P.; R. Flick; and D. R. Cayan (2003). Storminess variability along the California coast: 1858–2000. *J. Climate*, **16**, 982–993.

Bromirski, P.; D. Cayan; and R. Flick (2005). Wave spectral energy variability in the northeast Pacific. *J. Geophys. Res.*, **110**, C03005, doi: 10.1029/2004JC002398.

Caires, S.; V. Swail; and X. Wang (2006). Projection and analysis of extreme wave climate. *J. Climate*, **19**, 5581–5605, doi: 10.1175/JCLI3918.1.

CCSP (2008). *Weather and Climate Extremes in a Changing Climate. Regions of Focus: North America, Hawaii, Caribbean, and U.S. Pacific Islands*, a report by the U.S. Climate Change Science Program and the Subcommittee on Global Change Research [T. R. Karl, G. A. Meehl, C. D. Miller, S. J. Hassol, A. M. Waple, and W. L. Murray (eds.)], Department of Commerce, NOAA's National Climatic Data Center, Washington, D.C., 164 pp.

Chan, J. (2006). Comment on "Changes in tropical cyclone number, duration, and intensity in a warming climate." *Science*, **311**, 1713.

Chang, E. (2007). Assessing the increasing trend in Northern Atmosphere winter storm track activity using surface ship observations and a statistical storm track model. *J. Climate*, **20**, 5607–5628, doi: 10.1175/2007JCLI1596.1.

Chang, E.; and Y. Fu (2002). Interdecadal variations in Northern Hemisphere winter storm track intensity. *J. Climate*, **15**, 642–658.

Chang, E.; S. Lee; and K. Swanson (2002). Storm track dynamics. *J. Climate*, **15**, 2163–2182.

Chauvin, F.; J.-F. Royer; and M. Déqué (2006). Response of hurricane-type vortices to global warming as simulated by ARPEGE-climate at high resolution. *Climate Dyn.*, **27**, 377–399, doi: 10.1007/s00382-006-0135-7.

Church, J.; N. White; R. Coleman; K. Lambeck; and J. Mitrovica (2004). Estimates of the regional distribution of sea level rise over the 1950–2000 period. *J. Climate*, **17**, 2609–2625.

Church, J.; N. White; T. Aarup; W. Wilson; P. Woodworth; C. Domingues; J. Hunter; and K. Lambeck (2008). Understanding global sea levels: Past, present and future. *Sustain. Sci.*, **3**, 9–22, doi: 10.1007/s11625-008-0042-4.

Cieślikiewicz, W.; and B. Paplińska-Swerpel (2008). A 44-year hindcast of wind wave fields over the Baltic Sea. *Coastal Eng.*, 894–905, doi: 10.1016/j.coastaleng.2008.02.017.

Debernhard, J.; Ø. Sætra; and L. Roed (2002). Future wind, wave and storm surge climate in the Northern Seas. *Climate Res.*, **23**, 39–49.

Debernhard, J.; Ø. Sætra; and L. Roed (2008). Future wind, wave and storm surge climate in the Northern Seas: A revisit. *Tellus*, **A60**, 427–438, doi: 10.1111/j.1600-0870.2008.00312.x.

Domingues, C.; J. Church; N. White; P. Gleckler; S. Wijffels; P. Barker; and J. Dunn (2008). Improved estimates of upper-ocean warming and multi-decadal sea-level rise. *Nature*, **453**, 1090–1094, doi: 10.1038/nature07080.

Eichler, T.; and W. Higgins (2006). Climatology and ENSO-related variability of North American extra-tropical cyclone activity. *J. Climate*, **19**, 2076–2093.

Emanuel, K. (2005). Increasing destructiveness of tropical cyclones over the past 30 years. *Nature*, **436**, 686–688, doi: 10.1038/nature03906.

Emanuel, K. (2007). Environmental factors affecting tropical cyclone power dissipation. *J. Climate*, **20**, 5497–5509, doi: 10.1175/2007JCLI1571.1.

Emanuel, K.; and D. Nolan (2004). Tropical cyclone activity and global climate. In: *Proc. 26th Conf. on Hurricanes and Tropical Meteorology*. American Meteorological Society, Miami, FL.

Fischer-Bruns, I.; H. von Storch; F. González-Rouco; and E. Zorita (2005). Modelling the variability of midlatitude storm activity on decadal to century time scales. *Climate Dyn.*, doi: 10.1007/s00382-005-0036.1.

Flather, R.; and H. Khandker (1993). The storm surge problem and possible effects of sea level changes on coastal flooding in the Bay of Bengal. In: R. Warrick, E. Barrow, and T. Wigley (Eds.), *Climate and Sea Level Change: Observations, Projections and Implications.* Cambridge University Press, pp. 229–245.

Geng, Q.; and M. Sugi (2001). Variability of the North Atlantic cyclone activity in winter analyzed from NCEP-NCAR reanalysis data. *J. Climate*, **14**, 3863–3873.

Geng, Q.; and M. Sugi (2003). Possible change of extratropical cyclone activity due to enhanced greenhouse gases and sulfate aerosols–study with a high-resolution AGCM. *J. Climate*, **16**, 2262–2274.

Gibson, R.; P. Kålberg; and S. Uppala (1996). The ECMWF re-analysis (ERA) project. *ECMWF Newsl.*, **73**, 7–17.

Goldenberg, S.; C. Landsea; A. M. Núñez; and W. Gray (2001). The recent increase in Atlantic hurricane activity: Causes and implications. *Science*, **293**, 474–479.

Gorman, R.; K,. Bryan; and A. Laing (2003). Wave hindcast for the New Zealand region: Nearshore validation and coastal wave climate. *N.Z. J. Marine Freshwater Res.*, **37**, 567–588.

Grabemann, I.; and R. Weisse (2008). Climate change impact on extreme wave conditions in the North Sea: An ensemble study. *Ocean Dynamics*, **58**, 199–212, doi: 10.1007/s10236-008-0141-x.

Graham, N.; and H. Diaz (2001). Evidence for intensification of North Pacific winter cyclones since 1948. *Bull. Amer. Meteorol. Soc.*, **82**, 1869–1893.

Gray, W. (1968). A global view of the origin of tropical disturbances and storms. *Mon. Wea. Rev.*, **96**, 669–700.

Gulev, S.; and V. Grigorieva (2004). Last century changes in ocean wind wave height from global visual wave data. *Geophys. Res. Lett.*, **31**, L24601, doi: 10.1029/2004GL021032.

Gulev, S.; T. Jund;; and E. Ruprecht (2007). Estimation of the impact of sampling errors in the VOS observations on air–sea fluxes, Part II: Impact on trends and variability. *J. Climate*, **20**, 302–315.

Günther, H.; W. Rosenthal; M. Stawarz; J. Carretero; M. Gómez; I. Lozano, O. Serrano; and M. Reistad (1998). The wave climate of the Northeast Alantic over the period 1955–1994: The WASA wave hindcast. *Global Atmos. Oc. System*, **6**, 121–164.

Harnik, N.; and E. Chang (2003). Storm track variations as seen in radiosonde observations and reanalysis data. *J. Climate*, **16**, 480–495.

Hasegawa, A.; and S. Emori (2005). Tropical cyclones and associated precipitation over the western North Pacific: T106 atmospheric GCM simulation for present-day and doubled CO_2 climates. *Scientific Online Letters on the Atmosphere*, **1**, 145–148, doi: 10.2151/sola.2005-038.

Held, I. (1993). Large-scale dynamics and global warming. *Bull. Amer. Meteorol. Soc.*, **74**, 228–241.

Holgate, S.; S. Jevrejeva; P. Woodworth; and S. Brewer (2007). Comment on "A semi-empirical approach to projecting future sea level rise." *Science*, **317**, 1866, doi: 10.1126/science.1140942.

Holton, J. (1992). *An Introduction to Dynamic Meteorology*, Int. Geophys. Ser. Vol. 48. Academic Press, San Diego. ISBN 0-12-354355-X.

Houghton, J.; Y. Ding; D. Griggs; M. Noguer; P. van der Linden; X. Dai; K. Maskell; and C. Johnson (Eds.) (2001). *Climate Change 2001: The Scientific Basis. Contribution of Working Group I to the Third Assessment Report of the Intergovernmental Panel on Climate Change*. Cambridge University Press, Cambridge, U.K., 881 pp. ISBN 0521 01495 6.

Ishii, M.; M. Kimoto; K. Sakamoto; and S. Iwasaki (2006). Steric sea level changes estimated from historical ocean subsurface temperature and salinity analyses. *J. Oceanogr.*, **62**, 155–150.

Kaas, E.; U. Andersen; R. A. Flather; J. A. Williams; D. L. Blackman; P. Lionello; F. Dalan; E. Elvini; A. Nizzero; P. Malguzzi *et al.* (2001). *STOWASUS 2100: Regional Storm, Wave and Surge Scenarios for the 2100 Century*, Final Report ENV4-CT97-04989. Danish Meteorological Institute, Copenhagen, Denmark. Available at *http://web.dmi.dk/pub/STOWASUS-2100/Final/Synthesis.pdf*

Kaser, G.; J. Cogley; M. Dyorgerov; M. Meier; and A. Ohmura (2006). Mass balance of glaciers and ice caps: Consensus estimates for 1961–2004. *Geophys. Res. Lett.*, **33**, L19501, doi: 10.1029/2006GL027511.

Kauker, K.; and H. Langenberg (2000). Two models for the climate change related development of sea levels in the North Sea: A comparison. *Climate Res.*, **15**, 61–67.

Kharin, V.; and F. Zwiers (2005). Estimating extremes in transient climate change simulations. *J. Climate*, **18**, 1156–1172.

Klotzbach, P. (2006). Trends in global tropical cyclone activity over the past twenty years (1986–2005). *Geophys. Res. Lett.*, **33**, L10805, doi: 10.1029/2006GL025881.

Knight, J.; C. Folland; and A. Scaife (2006). Climate impacts of the Atlantic Multidecadal Oscillation. *Geophys. Res. Lett.*, **33**, L17706, doi: 10.1029/2006GL026242.

Knippertz, P.; U. Ulbrich; and P. Speth (2000). Changing cyclones and surface wind speeds over the North Atlantic and Europe in a transient GHG experiment. *Climate Res.*, **15**, 109–122.

Knutson, T.; and R. Tuleya (2004). Impact of CO_2-induced warming on simulated hurricane intensity and precipitation: Sensitivity to the choice of climate model and convective parameterization. *J. Climate*, **17**, 3477–3495.

Knutson, T.; and R. Tuleya (2008). Tropical cyclones and climate change: Revisiting recent studies at GFDL. In: H. Diaz and R. J. Murnane (Eds.), *Climate Extremes and Society*. Cambridge University Press, pp. 120–144. ISBN 978-0-521-87028-3.

Komar, P.; and J. Allan (2007). Higher waves along U.S. east coast linked to hurricanes. *EOS Trans.*, **88**, 301.

Komar, P.; and J. Allan (2008). Increasing hurricane-generated wave heights along the U.S. east coast and their climate controls. *J. Coastal Res.*, **24**, 479–488.

Kushnir, Y.; V. Cardone; J. Greenwood; and M. Cane (1997). The recent increase in North Atlantic wave heights. *J. Climate*, **10**, 2107–2113.

Lambert, S.; and J. Fyfe (2006). Changes in winter cyclone frequencies and strengths simulated in enhanced greenhouse warming experiments: Results from the models participating in the IPCC diagnostic exercise. *Climate Dyn.*, **26**, 713–728, doi: 10.1007/s00382-006-0110-3.

Landsea, C. (2005). Hurricanes and global warming. *Nature*, E11–E13, doi: 10.1038/nature 04477.

Landsea, C. (2007). Counting Atlantic tropical cyclones back to 1900. *EOS Trans*, **88**, 197–208, doi: 10.1029/2007EO180001.

Landsea, C.; C. Anderson; N. Charles; G. Clark; J. Dunion; J. Fernández-Patagás; P. Hungerford; C. Neumann; and M. Zimmer (2004). The Atlantic hurricane database reanalysis project: Documentation for 1851–1910 alterations and additions to the HURDAT data base. In: R. Murnane and K. Liu (Eds.), *Hurricanes and Typhoons: Past, Present and Future*. Columbia University Press, pp. 177–221.

Langenberg, H.; A. Pfizenmayer; H. von Storch; and J. Sündermann (1999). Storm-related sea level variations along the North Sea coast: Natural variability and anthropogenic change. *Continental Shelf Res.*, **19**, 821–842.

Lowe, J.; and J. Gregory (2005). The effects of climate change on storm surges around the United Kingdom. *Phil. Trans. R. Soc.*, **363**, 1313–1328, doi: 10.1098/rsta.2005.1570.

Mann, M.; and K. Emanuel (2006). Atlantic hurricane trends linked to climate change. *EOS Trans.*, **87**, 233–241.

Matulla, C.; and H. von Storch (2009). Changes in eastern Canadian storminess since 1880. *J. Climate*, submitted.

Matulla, C.; W. Schöner; H. Alexandersson; H. von Storch; and X. Wang (2008). European storminess: Late nineteenth century to present. *Climate Dyn.*, **31**, 1125–1130, doi: 10.1007/s00382-007-0333-y.

Maue, R. (2009). Northern Hemisphere tropical cyclone activity. *Geophys. Res. Lett.*, **36**, L05805, doi: 10.1029/208GL035946.

McCabe, G.; M. Clark; and M. Serreze (2001). Trends in Northern Hemisphere surface cyclone frequency and intensity. *J. Climate*, **14**, 2763–2768.

McDonald, R.; D. Bleaken; D. Cresswell; V. Pope; and C. Senior (2005). Tropical storms: Representation and diagnosis in climate models and the impacts of climate change. *Climate Dyn.*, **25**, 19–36, doi: 10.1007/s00382-004-0491-0.

McInnes, K.; and G. Hubbert (2003). A numerical modelling study of storm surges in Bass Strait. *Australian Meteorological Magazine*, **52**, 143–156.

McInnes, K.; K. Walsh; G. Hubbert; and T. Beer (2003). Impact of sea-level rise and storm surges on a coastal community. *Natural Hazards*, **30**, 187–207.

McInnes, K.; I. Macadam; G. Hubbert; D. Abbs; and J. Bathols (2005). *Climate Change in Eastern Victoria, Stage 2 Report: The Effect of Climate Change on Storm Surges*, a project undertaken for the Gippsland Coastal Board, Technical Report. CSIRO Marine and Atmospheric Research. Available at *http://www.cmar.csiro.au/e-print/open/mcinnes_2005b.pdf*

Meehl, G.; T. Stocker; W. Collins; P. Friedlingstein; A. Gaye; J. Gregory; A. Kitoh; R. Knutti; J. Murphy; A. Noda *et al.* (2007). Global climate projections. *Climate Change 2007: The Physical Science Basis. Contribution of Working Group I to the Fourth Assessment*

Sec. 5.7]   **References**   201

Report of the Intergovernmental Panel on Climate Change. Cambridge University Press, Cambridge, U.K. ISBN 978-0-521-88009-1.

Mitrovica, J.; M. Tamisiea; J. Davies; and G. Milne (2001). Recent mass balance of polar ice sheets inferred from patterns of global sea level change. *Nature*, **409**, 1026–1029.

Musić, S.; and S. Nicković (2008). 44-year wave hindcast for the Eastern Mediterranean. *Coastal Eng.*, 872–880, doi: 10.1016/j.coastaleng.2008.02.024.

Nakamura, H. (1992). Midwinter suppression of baroclinic wave activity in the Pacific. *J. Atmos. Sci.*, **49**, 1629–1642.

Oouchi, K.; J. Yoshimura; H. Yoshimura; R. Mizuta; S. Kusunoki; and A. Noda (2006). Tropical cyclone climatology in a global-warming climate as simulated in a 20 km–mesh global atmospheric model: Frequency and wind intensity analysis. *J. Meteorol. Soc. Japan*, **84**, 259–276, doi: 10.2151/jmsj.84.259.

Paciorek, C.; J. Risbey; V. Ventura; and R. Rosen (2002). Multiple indices of Northern Hemisphere cyclone activity, winters 1949–99. *J. Climate*, **15**, 1573–1590.

Peixoto, J.; and A. Oort (1992). *Physics of Climate.* American Institute of Physics. ISBN 0-88318-712-4.

Pinto, J.; U. Ulbrich; G. Leckebusch; T. Spangehl; M. Reyers; and S. Zacharias (2007). Changes in storm track and cyclone activity in three SRES ensemble experiments with the ECHAM5/MPI-OM1 GCM. *Climate Dyn.*, **29**, 195–210, doi: 10.1007/s00382-007-0230-4.

Pluess, A. (2004). Nichtlineare Wechselwirkung der Tide auf Änderungen des Meeresspiegels im Übergangsbereich Küste/Ästuar am Beispiel der Elbe. In: G. Gönnert, H. Grassl, D. Kellat, H. Kunz, B. Probst, H. von Storch, and J. Sündermann (Eds.), *Proceedings of Workshop Klimaänderung und Küstenschutz, November 29–30, Hamburg, Germany.* Available at *http://coast.gkss.de/staff/storch/pdf/kliky.proc.0411.pdf* [in German].

Pugh, D. (2004). *Changing Sea Levels: Effects of Tides, Weather and Climate.* Cambridge University Press. ISBN 9780521532181.

Rahmstorf, S. (2007). A semi-empirical approach to projecting future sea-level rise. *Science*, **315**, 368–370, doi: 10.1126/science.1135456.

Santer, B.; T. Wigley; P. Gleckler; C. Bonfils; M. Wehner; K. AchutaRao; T. Barnett; J. Boyle; W. Brüggemann; M. Fiorino *et al.* (2006). Forced and unforced ocean temperature changes in Atlantic and Pacific tropical cyclogenesis regions. *Proc. Natl. Academy Sci.*, **103**, 13905–13910, doi: 10.1073/pnas.0602861103.

Schmidt, H.; and H. von Storch (1993). German Bight storms analysed. *Nature*, **365**, 791.

Schmith, T.; S. Johansen; and P. Thejll (2007). Comment on "A semi-empirical approach to projecting future sea level rise." *Science*, **317**, 1866c, doi: 10.1126/science.1143286.

Serreze, M.; F. Barse; R. Barry; and J. Rogers (1997). Icelandic low cyclone activity: Climatological features, linkages with the NAO, and relationships with recent changes in the Northern Hemisphere circulation. *J. Climate*, **10**, 453–464.

Simmonds, I.; and A. Keay (2000). Variability of Southern Hemisphere extratropical cyclone behavior 1958–97. *J. Climate*, **13**, 550–561.

Sinclair, M.; and I. Watterson (1999). Objective assessment of extratropical weather systems in simulated climates. *J. Climate*, **12**, 3467–3485.

Solomon, S.; D. Qin; M. Manning; Z. Chen; M. Marquis; K. Averyt; M. Tignor; and H. Miller (Eds.) (2007). *Climate Change 2007: The Physical Science Basis. Contribution of Working Group I to the Fourth Assessment Report of the Intergovernmental Panel on Climate Change.* Cambridge University Press, Cambridge, U.K., 996 pp. ISBN 978-0-521-88009-1.

Stammer, D.; and S. Hüttemann (2008). Response of regional sea level to atmospheric pressure loading in climate change scenarios. *J. Climate*, **21**, 2093–2101, doi: 10.1175/2007JCLI1803.1.

Sterl, A.; and S. Caires (2005). Climatology, variability and extrema of ocean waves: The web-based KNMI/ERA-40 wave atlas. *Int. J. Climatol.*, 963–977, doi: 10.1002/joc.1175.

Sterl, A.; G. Komen; and P. Cotton (1998). Fifteen years of global wave hindcasts using winds from the European Centre for Medium-range Weather Forecasts reanalysis: Validating the reanalyzed winds and assessing wave climate. *J. Geophys. Res.*, **103**, 5477–5492.

Sterl, A.; R. Weisse; J. Lowe; and H. von Storch (2008). Storm climate. In: *Exploring High-end Climate Change Scenarios for Flood Protection of The Netherlands*, international scientific assessment carried out at the request of the Delta Committee, The Hague, The Netherlands.

Sugi, M.; A. Noda; and N. Sato (2002). Influence of global warming on tropical cyclone climatology: An experiment with the JMA global model. *J. Meteorol. Soc. Japan*, **80**, 249–272, doi: 10.2151/jmsj.80.249.

Trenberth, K.; P. Jones; P. Ambenje; R. Bojariu; D. Easterling; A. K. Tank; D. Parker; F. Rahimzadeh; J. Renwick; M. Rusticucci *et al.* (2007). *Climate Change 2007: The Physical Science Basis. Contribution of Working Group I to the Fourth Assessment Report of the Intergovernmental Panel on Climate Change*. Cambridge University Press, Cambridge, U.K. ISBN 978-0-521-88009-1.

Tsutsui, J. (2002). Implications of anthropogenic climate change for tropical cyclone activity: A case study with the NCAR/CCM2. *J. Meteorol. Soc. Japan*, **80**, 45–65, doi: 10.2151/jmsj.80.45.

Unnikrishnan, A.; K. R. Kumar; E. S. Fernandes; G. Michael; and S. Patwardhan (2006). Sea level changes along the Indian coast: Observations and projections. *Current Sci.*, **90**, 362–368. Available at *http://www.ias.ac.in/currsci/feb102006/362.pdf*

Vecchi, G.; and B. Soden (2007). Inceased tropical Atlantic wind shear in model projections of global warming. *Geophys. Res. Lett.*, **34**, L08702, doi: 10.1029/2006GL028905.

Vermeersen, L.; and H. Schotman (2009). Constraints of glacial isostatic adjustment from GOCE and sea level data. *Pure Appl. Geophys.*, in press, doi: 10.1007/s00024-004-0503-3.

von Storch, H.; and H. Reichardt (1997). A scenario of storm surge statistics for the German Bight at the expected time of doubled atmospheric carbon dioxide concentrations. *J. Climate*, **10**, 2653–2662.

von Storch, H.; and F. Zwiers (1999). *Statistical Analysis in Climate Research*. Cambridge University Press, New York, 494 pp.

von Storch, H.; G. Gönnert; M. Meine; and K. Woth (2008a). Storm surges: An option for Hamburg, Germany, to mitigate expected future aggravation of risk. *Environ. Sci Policy*, doi: 10.1016/j.envsci.2008.08.003.

von Storch, H.; E. Zorita; and J. González-Rouco (2008b). Relationship between global mean sea-level and global mean temperature in a climate simulation of the past millennium. *Ocean Dynamics*, **58**, 227–236, doi: 10.1007/s10236-008-0142-9.

Wang, X.; and V. Swail (2001). Changes of extreme wave heights in the Northern Hemisphere oceans and related atmospheric circulation regimes. *J. Climate*, **14**, 2204–2221.

Wang, X.; and V. Swail (2006a). Climate change signal and uncertainty in projections of ocean wave heights. *Climate Dyn.*, **26**, 109–126, doi: 10.1007/s00382-005-0080-x.

Wang, X.; and V. Swail (2006b). Historical and possible future changes of wave heights in northern hemisphere oceans. In: W. Perrie (Ed.), *Atmosphere Ocean Interactions*, Vol. 2. Wessex Institute of Technology Press, Ashurst, U.K., pp. 185–219.

Wang, X.; F. Zwiers; and V. Swail (2004). North Atlantic ocean wave climate change scenarios for the twenty-first century. *J. Climate*, **17**, 2368–2383.

Wang, X.; V. Swail; F. Zwiers; X. Zhang; and Y. Feng (2009). Detection of external influence on trends of atmospheric storminess and northern ocean wave heights. *Climate Dyn.*, **32**, 189–203, doi: 10.1007/s00382-008-0442-2.

WASA-Group (1998). Changing waves and storms in the Northeast Atlantic? *Bull. Amer. Meteorol. Soc.*, **79**, 741–760.

Watterson, I. (2005). Simulated changes due to global warming in the variability of precipitation, and their interpretation using a gamma-distributed stochastic model. *Adv. Water Res.*, **28**, 1368–1381.

Weisse, R.; and H. Günther (2007). Wave climate and long-term changes for the southern North Sea obtained from a high-resolution hindcast 1958–2002. *Ocean Dynamics*, **57**, 161–172, doi: 10.1007/s10236-006-0094-x.

Willis, J.; D. Roemmich; and B. Cornuelle (2004). Interannual variability in upper ocean heat content, temperature, and thermosteric expansion on global scales. *J. Geophys. Res.*, **109**, C12036, doi: 10.1029/2003JC002260.

WMO (2006). *Atmospheric Research and Environment Programme: Statement on Tropical Cyclones and Climate Change*. Available at *http://www.wmo.int/pages/prog/arep/tmrp/documents/iwtc_statement.pdf*

Woodworth, P.; and D. Blackman (2002). Changes in extreme high waters at Liverpool since 1768. *Int. J. Climatol.*, **22**, 697–714.

Woodworth, P.; and D. Blackman (2004). Evidence for systematic changes in extreme high waters since the mid-1970s. *J. Climate*, **17**, 1190–1197.

Woth, K. (2005). North Sea storm surge statistics based on projections in a warmer climate: How important are the driving GCM and the chosen emission scenario? *Geophys. Res. Lett.*, **32**, L22708, doi: 10.1029/2005GL023762.

Woth, K.; R. Weisse; and H. von Storch (2006). Climate change and North Sea storm surge extremes: An ensemble study of storm surge extremes expected in a changed climate projected by four different regional climate models. *Ocean Dynamics*, **56**, 3–15, doi: 10.1007/s10236-005-0024-3, online 2005.

Yin, J. (2005). A consistent poleward shift of the storm tracks in simulations of 21st century climate. *Geophys. Res. Lett.*, **32**, L18701, doi: 10.1029/2005GL023684.

Yoshimura, J.; M. Sugi; and A. Noda (2006). Influence of greenhouse gas warming on tropical cyclone frequency. *J. Meteorol. Soc. Japan*, **84**, 405–428.

Zhang, K.; B. Douglas; and S. Leatherman (2000). Twentieth-century storm activity along the U.S. east coast. *J. Climate*, **13**, 1748–1761.

Appendices

A.1 SCALE ANALYSIS

The concepts of *scales* and of *scale analysis* play a fundamental role in geoscience (e.g., in meteorology or oceanography). The term *scale* generally refers to the *typical* or *characteristic* dimensions of a process or phenomenon, such as the length or the spatial scale of, for example, thunderstorms, or the time scale (i.e., the typical duration or lifetime) of, for instance, mid-latitude cyclones. *Scale analysis* is a method that makes use of the non-dimensional form of equations to determine which terms are relevant for a particular phenomenon or for a situation, such that smaller terms can be neglected (Glickman, 2000). The result of scale analysis is a simplified set of equations that can be solved more easily.

The spectrum of motions in the atmosphere and oceans covers a wide range of space and time scales. In the atmosphere, the types of motion range from small-scale turbulence, local wind systems, synoptic scale motions within mid-latitude high-pressure and low-pressure systems, to global atmospheric circulation. These different types of motions are all governed by the same set of fundamental equations. Within this set of equations, the different scales are, more or less strongly, coupled via non-linear terms.

The types of motion can be characterized by their *characteristic* spatial extent and their average lifetime T (i.e., their space and time *scale*). For spatial scales, a separation is usually made between horizontal (L) and vertical scales (D) as well. Table A.1 presents some examples of atmospheric phenomena and their characteristic space and time scales. According to their spatial scale, atmospheric motions are often classified as small-scale, convective-scale, meso-scale, or large-scale motions (Table A.2).

When a particular scale of motion is considered, it is appropriate to simplify the governing equations and retain only those terms that are significant for description of the considered phenomenon. The technique to do so is called scale analysis

206 **Appendices**

Table A.1. Characteristic horizontal scales L, vertical scales D, and time scales T of some types of atmospheric motion (after Pichler, 1997).

Type of motion	L (m)	D (m)	T (s)
Small-scale turbulence	$10^{-2} \cdots 10^2$	$10^{-2} \cdots 10^2$	$1 \cdots 10^1$
Cumulus convection	10^3	10^3	10^2
Cumulonimbus convection	10^4	10^4	10^3
Fronts, squall lines	10^5	10^4	10^4
Synoptic cyclones	10^6	10^4	10^5
Planetary waves	10^7	10^4	10^6

(e.g., Charney, 1948; Pichler, 1997; Holton, 1992). It provides objective criteria to derive simplified equations that are appropriate for describing the problem considered. Some care is needed, however, since interaction among motions that have different scales may occur.

The principal approach to carrying out scale analysis is the following (e.g., Jacobson, 1968; Pichler, 1997): given, in general, an equation of the form

$$\sum_i X_i = 0. \tag{A.1.1}$$

The equation can formally be written as

$$\sum_i \hat{X}_i X_i^* = 0, \tag{A.1.2}$$

where the $\hat{X}_i = \text{magn}(X_i)$ represent typical (expected) values of the *magnitude* of field variables X_i; and the X_i^* are dimensionless terms with magnitude *one*.[1] The ratio between the magnitude of the different terms $\hat{X}_i \hat{X}_j^{-1}$ in (A.1.2) then represents a measure on how important the different terms are relative to each other. Comparing the different ratios allows identification of the relevant terms for the problem considered and simplifying the equations by removing those terms that have only little influence.

In meteorology and oceanography specific names are used for some of these ratios. For example, the ratio between the characteristic magnitudes of inertial and Coriolis force is referred to as the *Rossby number* after the Swedish meteorologist C. G. Rossby (1898–1957). A small Rossby number signifies a system that is strongly affected by Coriolis forces, while a large Rossby number characterizes a system in which inertial and centrifugal forces dominate. A small Rossby number is required for geostrophic approximation to hold (Appendix A.2). For a more detailed descrip-

---

[1] $\text{magn}(X_i^*) \cong 1$.

Table A.2. Classification of atmospheric motions according to their horizontal scale L.

Scale	L (m)
Small scale	$< 10^3$
Convective scale	$10^3 \cdots 10^4$
Meso scale	$10^4 \cdots 10^6$
Large scale	$> 10^6$

tion of scales and scale analysis we refer the reader to existing textbooks such as Holton (1992), Apel (1987), or Pichler (1997).

A.2 GEOSTROPHIC WIND

The *geostrophic wind* represents a fundamental approximation in meteorology. It is fully determined by the atmospheric pressure field and describes, in many cases, true wind conditions at first approximation. For example, storm indices based on geostrophic wind speeds are frequently used to describe long-term variations in storm climate (e.g., Schmidt and von Storch, 1993; Alexandersson *et al.*, 1998; Matulla *et al.*, 2008) (see also Section 4.3). To derive the geostrophic wind equation, we start with Newton's Second Law of Motion which, in an inertial reference frame, can formally be expressed by

$$\frac{d\mathbf{v}}{dt} = \sum_i \mathbf{f}_i, \qquad (A.2.1)$$

where the left-hand side represents the rate of change of the three-dimensional velocity vector $\mathbf{v}$ (acceleration); and the right-hand side represents the sum of all forces per unit mass acting on an air parcel.

In the atmosphere, the fundamental forces acting on an air parcel in an inertial reference frame are the pressure gradient force $\mathbf{P}$, the gravitational force $\mathbf{G}$, and the frictional forces $\mathbf{F}$. The right-hand side of (A.2.1) may thus be expressed as

$$\sum_i \mathbf{f}_i = \mathbf{P} + \mathbf{G} + \mathbf{F}, \qquad (A.2.2)$$

or, in more detail and with minor approximations for gravitational force, as

$$\sum_i \mathbf{f}_i = -\frac{1}{\rho}\nabla p + g\mathbf{k} + \mathbf{F}, \qquad (A.2.3)$$

where ρ denotes air density; p is the three-dimensional pressure field; g is acceleration due to gravity; and $\mathbf{k}$ represents a unit vector perpendicular to the Earth's surface. For details on how to derive expressions for the fundamental forces in (A.2.3) we refer the reader to existing textbooks, such as Holton (1992) or Pichler (1997).

Newton's Second Law as expressed in (A.2.3) is valid only in an inertial reference frame. As the atmosphere and Earth are rotating, the governing equations are usually expressed in a rotating (i.e., a non-inertial) reference frame. In such a reference frame, additional forces due to the non-uniform relative motion of two reference frames have to be accounted for. Such fictitious forces are usually called *apparent* or *pseudo* forces. They are not the result of physical interactions, but merely result from

208   **Appendices**

acceleration of the non-inertial reference frame itself. In the atmosphere, the significant pseudo-forces are the centrifugal and Coriolis forces.

Centrifugal force accounts for the effect rotation has on an object at rest in a rotating reference frame. When the object is not at rest in the rotating reference frame, an additional pseudo-force referred to as the *Coriolis force* emerges. As the atmosphere and the oceans are not at rest relative to the rotating Earth, the Coriolis force represents an important pseudo-force in meteorology and oceanography. Its effect can be demonstrated by means of the following Gedankenexperiment[2]: suppose that the Earth's atmosphere is at rest relative to Earth's surface; that is, looked upon from an inertial reference frame, the atmosphere is rotating at the same angular speed as the solid Earth. In other words, each air parcel within the atmosphere is moving eastward at speed $\omega \cos \phi a^2$, where ω is the Earth's angular speed of rotation, a is the radius of the Earth, and ϕ represents latitude. Thus the eastward-traveling component of each air parcel depends on latitude, with higher speeds occurring near the Equator and lower speeds near the poles. Suppose now that an air parcel near the Equator is set into poleward motion. As of now, it is no longer at rest within the rotating reference frame. As the air parcel encounters higher latitudes, the air masses present at these latitudes will have lower eastward velocities. Thus, the air parcel set into poleward motion appears to become faster than its surroundings and, looked upon from a rotating reference frame, the path of the air parcel appears to be deflected relative to a straight line. The apparent or pseudo-force responsible for this deflection is called the Coriolis force. The Coriolis force is perpendicular to the velocity vector itself and can only change direction, but not the speed of the motion.

Taking pseudo-forces into account, the equation of motion (A.2.1) in a rotating reference frame reads (Holton, 1992)

$$\left. \begin{array}{l} \dfrac{du}{dt} - \dfrac{uv \tan \phi}{a} + \dfrac{uw}{a} = -\dfrac{1}{\rho}\dfrac{\partial p}{\partial x} + fv - f^*w + F_x \\[2ex] \dfrac{dv}{dt} + \dfrac{u^2 \tan \phi}{a} + \dfrac{uw}{a} = -\dfrac{1}{\rho}\dfrac{\partial p}{\partial y} - fu + F_y \\[2ex] \dfrac{dw}{dt} + \dfrac{u^2 + v^2}{a} = -\dfrac{1}{\rho}\dfrac{\partial p}{\partial z} - g + f^*u + F_z, \end{array} \right\} \qquad \text{(A.2.4)}$$

where u, v, and w represent the zonal (eastward), meridional (northward), and vertical (upward) velocity components; F_x, F_y, and F_z denote the corresponding components of the frictional force and

$$\left. \begin{array}{l} f = 2\omega \sin \phi, \\[1ex] f^* = 2\omega \cos \phi \end{array} \right\} \qquad \text{(A.2.5)}$$

are known as Coriolis parameters.

Applying scale analysis (see Appendix A.1) with characteristic values for synoptic-scale motions in mid-latitudes (Table A.3) reveals that the magnitudes of

---

[2] An experiment carried out in thought only.

Appendices 209

Table A.3. Characteristic scales for synoptic-scale motions at mid-latitudes.

Horizontal velocity scale	U	10	$\mathrm{m\,s^{-1}}$
Vertical velocity scale	V	10^{-2}	$\mathrm{m\,s^{-1}}$
Horizontal length scale	L	10^{6}	m
Vertical length scale	H	10^{4}	m
Scale of horizontal pressure fluctuations	$\delta p/\rho$	10^{3}	$\mathrm{m^2\,s^{-2}}$
Time scale	L/U	10^{5}	s

the additional acceleration terms on the left-hand side of (A.2.4) are small compared with all other terms in the equations. For synoptic-scale motions, (A.2.4) may hence be reduced and expressed more conveniently as

$$\frac{d\mathbf{v}}{dt} + \mathbf{k} \times f\mathbf{v}_h + \mathbf{i}wf^* - \mathbf{k}uf^* = -\frac{1}{\rho} \times \nabla p + \mathbf{k}g + \mathbf{F}, \tag{A.2.6}$$

where $\mathbf{v}_h = [u, v, 0]$ denotes the horizontal wind vector.

Equation (A.2.6) may be reconsidered using the technique of scale analysis and the arguments for synoptic scales at mid-latitudes. It appears that, as far as the horizontal components of the equation are concerned, the Coriolis force $\mathbf{k} \times f\mathbf{v}_h$ is approximately in balance with the pressure gradient force and that the typical magnitude of all other terms is at least one order of magnitude smaller. Equation (A.2.6) can thus be reduced to

$$\mathbf{v_g} = \mathbf{k} \times \frac{1}{\rho f} \nabla p. \tag{A.2.7}$$

Equation (A.2.7) is known as the geostrophic wind relation and we have introduced the subscript g to denote the wind vector derived from the geostrophic approximation. The geostrophic wind relation represents a fundamental approximation in meteorology. It states that there exists a diagnostic relation between atmospheric pressure and the horizontal wind field and that the geostrophic wind vector can be determined diagnostically from the pressure field without any additional information on other variables. From (A.2.7) it can be further inferred that the geostrophic wind speed is proportional to the atmospheric pressure gradient and that the geostrophic wind direction is parallel to the isobars.

In many cases, the geostrophic wind represents a rather good approximation of the true wind conditions for synoptic-scale motions away from the surface. To demonstrate this, we apply the operation $\mathbf{k} \times \cdots$ to both sides of (A.2.6):

$$\mathbf{k} \times \frac{d\mathbf{v}}{dt} + \mathbf{k} \times (\mathbf{k} \times f\mathbf{v}_h) + \mathbf{k} \times \mathbf{i}wf^* - \mathbf{k} \times \mathbf{k}uf^* = -\frac{1}{\rho}\mathbf{k} \times \nabla p + \mathbf{k} \times \mathbf{k}g + \mathbf{k} \times \mathbf{F}.$$

$$\tag{A.2.8}$$

210   **Appendices**

Rearranging (A.2.8) yields

$$\mathbf{v}_h = \frac{\mathbf{k}}{\rho f} \times \nabla p + \frac{\mathbf{k}}{f} \times \frac{d\mathbf{v}}{dt} - \frac{\mathbf{k}}{f} \times \mathbf{F} + \mathbf{j}w \cot \phi. \tag{A.2.9}$$

The first term on the right-hand side of (A.2.9) represents the geostrophic wind while the other terms denote deviations from the geostrophic balance. They are caused by individual acceleration ($\mathbf{v}_{ag}^a$), friction ($\mathbf{v}_{ag}^f$), and latitudinal effects ($\mathbf{v}_{ag}^l$). Hence the horizontal wind vector may be expressed as

$$\mathbf{v}_h = \mathbf{v}_g + \mathbf{v}_{ag}^a + \mathbf{v}_{ag}^f + \mathbf{v}_{ag}^l. \tag{A.2.10}$$

Scale analysis of (A.2.10) reveals that, for synoptic-scale motions and away from the surface, the magnitude of the geostrophic wind is in the order of 10^1 m s^{-1} while the magnitudes of ageostrophic components vary between 10^0 m s^{-1} and 10^{-2} m s^{-1}. Hence, the geostrophic wind accounts for about 80% to 90% of the horizontal wind vector away from the surface (i.e., when friction is small). In other words, on synoptic scales a very large fraction of horizontal wind in the free atmosphere is caused by atmospheric pressure gradients and can be approximated well by the geostrophic wind relation.

A.3 GEOPOTENTIAL HEIGHT AND PRESSURE AS VERTICAL COORDINATES

Gravity $\mathbf{g}$ can be expressed in terms of a potential function Φ called the *geopotential*

$$d\Phi = g \, dz. \tag{A.3.1}$$

If the value of the geopotential is set to zero at mean sea level, then $\Phi(z)$ represents the work required to raise an air parcel of unit mass to height z from mean sea level.

Because of inhomogeneities in the solid Earth, gravity is not a constant but depends on latitude ϕ and longitude λ. Gravity also changes with height and in consequence

$$g = g(\phi, \lambda, z). \tag{A.3.2}$$

Meteorologists often prefer to replace $g \, dz$ in (A.3.1) with $g_n \, dh$

$$g \, dz = g_n \, dh, \tag{A.3.3}$$

where $g_n = 9.80665$ m s^{-2} represents the global average of gravity at mean sea level; and h is called the *geopotential height*. The unit of h is called the *geopotential meter* or *gpm*. In other words, to move an air parcel within the Earth's gravity field over a distance of $\Delta h = 1$ gpm requires the same amount of work as to move it over a vertical distance of 1 meter under average conditions

$$\Delta z = \frac{g_n}{g} \Delta h. \tag{A.3.4}$$

At the equator, $g < g_n$ and the vertical distance over which to move an air parcel for 1 gpm is larger than 1 meter. Near the poles the situation is reversed.

From the vertical component of the equation of motion (A.2.4) it can be inferred that in the absence of atmospheric motions gravity is balanced exactly by the vertical component of the pressure gradient force

$$\frac{\partial p}{\partial z} = -\rho g. \tag{A.3.5}$$

Equation (A.3.5) is called the *hydrostatic balance* and represents a good approximation for large-scale atmospheric conditions. From (A.3.5) it follows that there exists a single-valued monotonic relation between pressure and height. Any such single-valued monotonic function of pressure or height may be used as an independent vertical coordinate. In numerical models of atmospheric pressure, or pressure normalized by surface pressure, they are often used. The latter are usually referred to as σ-coordinates with

$$\sigma = \frac{p(x, y, z, t)}{p(x, y, z = 0, t)} \tag{A.3.6}$$

and $\sigma = 1$ at the surface.

A.4 THERMAL WIND

Using the definition of the geopotential (A.3.1), the geostrophic wind relation (A.2.7) can be rewritten as

$$\mathbf{v_g} = \frac{\mathbf{k}}{f} \times \nabla_p \Phi, \tag{A.4.1}$$

where $\nabla_p \Phi$ represents the gradient of the geopotential along surfaces of constant pressure. Using the ideal gas law $p = \rho R T$ with T being temperature and $R = 287 \, \mathrm{J\,kg^{-1}\,K^{-1}}$ being the gas constant for dry air, the hydrostatic equation (A.3.5) can be expressed as

$$\frac{\partial \Phi}{\partial p} = -\frac{RT}{p}. \tag{A.4.2}$$

Differentiating (A.4.1) with respect to pressure and applying (A.4.2) yields an equation for the vertical shear of the geostrophic wind vector

$$p \frac{\partial \mathbf{v_g}}{\partial p} = \frac{\partial \mathbf{v_g}}{\partial \ln p} = -\frac{\mathbf{k} R}{f} \times \nabla_p T. \tag{A.4.3}$$

As shear only depends on the temperature gradient along surfaces of constant pressure, (A.4.3) is called the *thermal wind* equation and the shear of the geostrophic wind is referred to as the *thermal wind*

$$\mathbf{v_{th}} = -\frac{\mathbf{k} R}{f} \times \nabla_p T. \tag{A.4.4}$$

The thermal wind equation provides a valuable diagnostic tool to check analyses of observed wind and temperature fields. The thermal wind is directed parallel to the isotherms with warm air to the right and cold air to the left if one looks into the

212 **Appendices**

direction the thermal wind is blowing. The thermal wind equation can thus be used to estimate temperature advection even from a single vertical wind profile or sounding. When the geostrophic wind turns counterclockwise with height, the advection of cold air can be inferred and *vice versa*.

A.5 LIST OF SYMBOLS

Only principal symbols are listed. Symbols formed by adding primes, overbars, etc. are not listed. Boldface type indicates vector quantities. Some symbols are used to represent more than one variable in order to maintain traditional usage. Where symbols have more than one meaning, this is indicated in the text.

a	Mean radius of the Earth (6,370 km)
β	Variation of Coriolis parameter f with latitude
c_d	Drag coefficient
η	Absolute vorticity
f	Coriolis parameter ($2\omega \sin \phi$)
F	Wave spectrum
ϕ	Latitude
Φ	Geopotential
g	Gravitational acceleration ($9.806 \, \mathrm{m \, s^{-2}}$)
GP	Genesis potential index
H	Significant wave height
h	(1) Individual wave height
	(2) Sea surface elevation
	(3) Depth, height, or layer thickness
K_h	Horizontal eddy viscosity
K_v	Vertical eddy viscosity
$\mathbf{k}$	(1) Unit vector perpendicular to the Earth's surface
	(2) Wavenumber vector
λ	Longitude
∇	Vector gradient operator
$\nabla \cdot$	Divergence operator
ω	Frequency
p	Pressure
q	Specific humidity
R	Gas constant for air ($287 \, \mathrm{J \, kg^{-1} \, K^{-1}}$)
RF	Relative humidity
ρ	Density
s	Surge height
S	Salinity
σ	Standard deviation
t	(1) Time
	(2) Tidal component of sea surface height

T	Temperature
Θ	Directional spectrum
u	Zonal velocity component
u_{10}	Wind speed at 10 m height
$\mathbf{u}$	Velocity vector
v	Meridional velocity component
$\mathbf{v}_g$	Geostrophic wind vector
$\mathbf{v}_{th}$	Thermal wind vector
z	Vertical coordinate
ζ	Relative vorticity

A.6 REFERENCES

Alexandersson, H.; T. Schmith; K. Iden; and H. Tuomenvirta (1998). Long-term variations of the storm climate over NW Europe. *Global Atmos. Oc. System*, **6**, 97–120.

Apel, J. (1987). *Principles of Ocean Physics*. Academic Press. ISBN 0-120-58866-8.

Charney, J. (1948). On the scale of atmospheric motions. *Geof. Publ.*, **17**.

Glickman, T. (Ed.) (2000). *Glossary of Meteorology*. American Meteorological Society, Boston, MA, Second Edition.

Holton, J. (1992). *An Introduction to Dynamic Meteorology*, Int. Geophys. Ser. Vol. 48. Academic Press, San Diego, CA. ISBN 0-123-54355-X.

Jacobson, J. (1968). *Einige Betrachtungen zur Verwendung der Scale Analyse in der dynamischen Meteorologie*, Technical Report. Veröff. d. Inst. f. Theoret. Meteorol. der FU Berlin [in German].

Matulla, C.; W. Schöner; H. Alexandersson; H. von Storch; and X. Wang (2008). European storminess: Late nineteenth century to present. *Climate Dyn.*, **31**, 1125–1130, doi: 10.1007/s00382-007-0333-y.

Pichler, H. (1997). *Dynamik der Atmosphäre*. Spektrum Akademie Verlag, Heidelberg, Germany [in German]. ISBN 3-827-40134-8.

Schmidt, H.; and H. von Storch (1993). German Bight storms anaysed. *Nature*, **365**, 791.

Index

accumulated cyclone energy (ACE), 174
added value, 95, 146–149
analysis, 121, 129
apparent forces, 207–208
attribution, 152–154

barocline, 27, 28
barotrop, 27, 28
Big Brother Experiment, 93

climate
  definition of, 1–3
  feedback of regional on planetary, 20–22
  interplay between scales, 16–18
  regional controlled by planetary, 18–20
climate change scenarios, *see* scenarios
climate models, 82–83
  global, 87–90
  regional, *see* regional climate models
climate system, 3–4
climate variability, 12–22
  externally forced, 12–16
  internally driven, 12–16
  stochastic climate model, *see* stochastic
    climate model
consensus, *see* scientific consensus
consistent data sets, 121
Coriolis force, *see* pseudo forces

data quality, problems, 114–123

detection, 152–153
  fingerprints, 154
  trends, 154–156
detection and attribution, 16, 151–156
diagnostic variables, 84–85
discretization
  finite differences, 85
  spectral approach, 86
downscaling, 18, 144–149
  added value, 146–149
  dynamical, 91, 145
  statistical, 146
  statistical–dynamical, 146

Ekman transport, 37, 62
equations
  Navier–Stokes, 103
  geostrophic wind, 209
  horizontal wind, 210
  hydrostatic balance, 211
  primitive equations, 83
  shallow water, 105–106
  thermal wind, 211
  wave action density balance, 99
  wave energy, 98
  wave height, fully developed sea, 46
  wave period, fully developed sea, 46
ERA-40 reanalysis, 140
exceedance probability, 50

216 Index

extra-tropical cyclones, *see* mid-latitude
    storms
extra-tropical storms, *see* mid-latitude
    storms
extreme sea levels, *see* storm surges

fair weather bias, 181
Ferrel cell, *see* general circulation
freak waves, *see* rogue waves
friction velocity, 100
future changes
    mean sea level, 189–191
    mid-latitude storms, 171–173
      mechanisms, 171
    sea state, 184–186
    storm surges, 193–194
    storm tracks, 171–173
      mechanisms, 171
    tides, 194–195
    tropical cyclones, 176–179
      approaches for analysis, 177–178
    wind waves, 184–186

general circulation
    atmosphere, 4–8, 28, 29
      Ferrel cell, 7
      Hadley cell, 7
      intertropical convergence zone, 7
      mean westwind circulation, 28, 29
      mid-latitude storms, 29, 31
      planetary waves, 28, 29
      Rossby waves, 31
      storm track, 32
      zonal mean flow, 28, 29
    oceans, 812
      density-driven circulation, 11–12
      meridional overturning circulation, 8,
        11–12
      thermohaline circulation, 8, 11–12
      western boundary currents, 9
      wind-driven circulation, 9–11
general circulation models, 83–87
genesis potential index (GP), 38, 178
geopotential height, 210–211
geopotential meter, 210
geostrophic wind, 207–210
gravity waves, *see* sea state

Hadley cell, *see* general circulation

heat transport, *see* meridional heat
    transport
Heney forcing, *see* nudging
hindcast, *see* reconstructions
homogeneity, 115–118
    changing data availability, 118–120
    creeping inhomogeneity, 117–118
    sudden inhomogeneity, 116–117
horizontal wind vector, 207–210
hydrostatic balance, 211

ICOADS data, 123
inhomogeneity, *see* homogeneity
intertropical convergence zone, *see* general
    circulation

Mann–Kendall test, 154
marine climate, *see* climate
maximum potential intensity, 177
mean sea level, 69–71
    atmospheric pressure loading, 191
    budget, 189
    eustatic changes, 70
    future changes, 189–191
    isostatic adjustment, 70
    isostatic changes, 70
    past changes, 187–189
    regional changes, 190–191
    self-gravitational effects, 190
    steric changes, 70
    tectonic movements, 70
    thermal expansion, 70
meridional heat transport, 3, 12
meridional overturning circulation, *see*
    general circulation
mid-latitude cyclones, *see* mid-latitude
    storms
mid-latitude storms, 27–35
    examples, 33–35
      Baltic Sea 2005 (Gudrun), 33
      North America 1993, 33
      North Sea 1953, 33
      North Sea 1962, 33
      Perfect or Halloween Storm, 34

Index 217

future changes, 171–173
  mechanisms, 171
past changes, 166–170
models, 77
  climate, *see* climate models
  cognitive, 79
  constructive use of, 78, 79
  general cicrulation, *see* general circulation
    models
  purpose of, 78
  quasi-realistic, 78–82
  resolution and grid size, 86
  tide–surge, *see* tide–surge models
  validation, 78
  validation vs. verification, 80
  wave, *see* wave models

NCEP/NCAR reanalysis, 135–140

parameterizations, 22, 78, 86
  drag coefficient, 105
past changes
  mean sea level, 187–189
  mid-latitude storms, 166–170
  sea state, 180–184
  storm surges, 192–193
  storm tracks, 166–170
  tides, 194–195
  tropical cyclones, 173–176
  wind waves, 180–184
percentiles, 19
polar lows, 72
power dissipation index (PDI), 175
pressure as vertical coordinate, 210–211
primitive equations, 83
prognostic variables, 84–85
projections, *see* scenarios
proxy data, 124–129
  storm activity, 124, 169
    geostrophic wind, 124–126
    multi-station indices, 124–126
    single-station indices, 126–127
    water level indices, 127–129
pseudo forces, 207–208

reanalysis, 130, 166, 167
  ERA-15, 90, 140
  ERA-40, 89, 140

extra-tropical cyclones, 132
global, 129–142
JRA-25, 140
NCEP, 90, 135–140
output variables, categories, 136
regional, *see* regional reanalysis
trends derived from, 130–135
reconstructions, 81, 142–143, 181
regional climate models, 90–96
  added value, 95–96, 146–149
  Big Brother Experiment, 93
  downscaling, *see* downscaling
  intermittent divergence, 93
  nudging, 91
  spectral nudging, 94–95
  sponge zone, 91
regional reanalysis, 142–144
  CaRD10, 143
  coastDat, 144
  NARR, 143
  reconstructions, 142–143
regional sea level changes, 190–191
  atmospheric pressure loading, 191
  self-gravitational effects, 190
regionalization techniques, *see* downscaling
rogue waves, 51–57
  detected from SAR, 55
  Draupner platform, 54
  examples, 52
  mechanisms, 55–57
  New Year Wave, 54
Rossby number, 206
Rossby waves, 29, 31

scale analysis, 205–207
scales, 205–207
  atmospheric motions, 206
  spatial scale, 205
  synoptic-scale motions, 209
  time scale, 3, 205
scenarios, 81, 149–151
scientific consensus, 165
sea breeze, 19
sea level, *see* mean sea level
sea level budget, 189
sea state, 96
  dispersion relation, 97
  duration, 45
  fetch, 45

218 **Index**

sea state (*cont.*)
  freak waves, *see* rogue waves
  fully developed, 45, 101
  future changes, 184–186
  gravity waves, 97
  models of, *see* wave models
  past changes, 180–184
  peak frequency, 103
  peak period, 103
  Rayleigh distribution, 49–51
  rogue waves, *see* rogue waves
  significant wave height, *see* significant
    wave height
  swell, 46, 47
  T_{m01} period, 103
  T_{m02} period, 103
  wave spectrum, *see* wave spectrum
  wind sea, 46
  wind waves, *see* wind waves
  zero-downcrossing period, 103
significant wave height, 49, 50, 102
  fully developed sea, 46
spectral nudging, 94
stochastic climate model, 13–16
storm surges, 57–66
  examples, 62–66
    Baltic Sea, 64–66
    Baltic Sea 1872, 65
    Baltic Sea 2005, 65
    North Sea, 62–64
    North Sea 1962, 64
  external, 59, 63
  future changes, 193–194
  inverse barometric effect, 60–62
  past changes, 192–193
  response to wind forcing, 62
  tide–surge interaction, 64
storm track, 27–35
  future changes, 171–173
    mechanisms, 171
  past changes, 166–170
sustained wind speed, 35
swell, 46, 47

thermal wind, 211–212
thermocline, 11

thermohaline circulation, *see* general
    circulation
tide–surge interaction, 64
tide–surge models, 103-106
  performance, 106
tides, 66–69
  amphidromic points, 68
  diurnal, 67
  future changes, 194–195
  harmonic analysis, 68
  long-term components, 67
  past changes, 194195
  semidiurnal, 67
  spring–neap cycle, 67
  tidal resonance, 68
  tide–surge interaction, 64
tropical cyclones, 3544
  accumulated cyclone energy (ACE), 174
  best track data, 122
  classification, 35
  conditions for formation, 36–38
  distribution, 36
  examples, 39–44
    Denise, 42
    Hyacinthe, 42
    Irma, 42
    Kate, 39, 42
    Katrina, 39, 42
    Tracy, 41
    Wilma, 40
  future changes, 176–179
    approaches for analysis, 177–178
  genesis potential index (GP), 38, 178
  maximum potential intensity, 177
  past changes, 173–176
  power dissipation index (PDI), 175
  structure, 38–39
  sustained wind speed, 35
  wind shear, 37, 38, 178
tropical storms, *see* tropical cyclones

uncertainty, in climate change studies,
    165–166

validation, *see* models
variables, prognostic and diagnostic, 84–85
voluntary observing ships, 123, 181
  fair weather bias, 181

Index 219

wave action density, 98
wave energy
    action density, 98
wave energy, conservation of, 98–102
    dissipation, 100
    wave–wave interaction, 100
    wind input, 99
wave models, 96–103
    first and second generation, 101
    third generation, 101
wave spectrum, 96–98
    directional spectrum, 98
    energy spectrum, 98
    frequency direction spectrum, 98
    frequency spectrum, 98
    power spectrum, 98
    two-dimensional wavenumber spectrum,
        97
    variance spectrum, 98
    wavenumber direction spectrum, 97
weather analyses, 121–123
western boundary currents, *see* general
        circulation

wind sea, 46
wind waves, 44–57
    capillary waves, 44
    cross sea, 47
    duration, 45
    fair weather bias, 181
    fetch, 45
    freak waves, *see* rogue waves
    fully developed, 45, 101
    future changes, 184–186
    generation, 44–48
        duration limited, 46
        fetch limited, 46
    gravity waves, 44, 97
    past changes, 180–184
    phase speed, smallest possible, 44
    rogue waves, *see* rogue waves
    shoaling, 47
    significant wave height, *see* significant
        wave height
    swell, 46, 47
    voluntary observing ships, 181
    wind sea, 46

Printing: Mercedes-Druck, Berlin
Binding: Stein+Lehmann, Berlin